£39-50

76416

S
V

Errata for *Plastic Design to BS 5950*

Page 7, Fig. 1.10 should be as below:

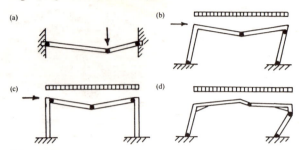

Page 11, Example 1.1(a), third line of calculation should read:

$$\therefore \quad \bar{y} = 29.58 \, \text{mm}$$

Page 62, Fig. 2.62, equation within diagram labelling should read:

$$\left(\frac{x}{11.25} + 0.5 \right)$$

Page 80, calculation immediately above Fig. 2.83 should read:

$$4.5\theta = \frac{9y}{5}\phi \quad \text{giving} \quad \theta = \frac{2y}{5}\phi$$

Page 96, Section 5.7.3.3, Clause (e), calculation (2) should read:

$$(2) \quad \text{when } 5.75 \le \lambda_{cr} < 20: \qquad \lambda_p \ge \frac{0.95\lambda_{cr}}{\lambda_{cr} - 1}$$

Page 113, Table 3.1, right-hand column, 6th paragraph should read:

Simplified version of Ref. 3.15

Page 118, 17 lines up from base of page, should read:

analysis which was derived by Horne.[3.15] It has been verified by

Page 119, Fig. 3.24, Reference citation within diagram labelling should read:

From formula of Ref. 3.15

Page 124, numbered list point (3), citation should read:

Reference 3.15

Page 174, seventh line of calculations should read:

$$\therefore \quad S = 49.2(39.53 - 19.865) + 24.946(0.20 + 19.798)$$

Page 182, sixth line of calculations should read:

$$B = \text{width of section} = 142.4 \, \text{mm}$$

Page 213, Table 6.3, first section should read:

	A	B	C
4th storey	162	324	162
	−1		+1
13/20	70	12	70
	231	336	233

PLASTIC DESIGN
TO BS 5950

J.M. Davies, *DSc, PhD, FICE, FIStructE*

B.A. Brown, *MSc, FIStructE, MASCE*

The Steel Construction Institute

Blackwell
Science

Copyright © 1996 The Steel Construction Institute

Blackwell Science Ltd
Editorial Offices:
Osney Mead, Oxford OX2 0EL
25 John Street, London WC1N 2BL
23 Ainslie Place, Edinburgh EH3 6AJ
238 Main Street, Cambridge,
  Massachusetts 02142, USA
54 University Street, Carlton
  Victoria 3053, Australia

Other Editorial Offices:
Arnette Blackwell SA
  224, Boulevard Saint Germain
  75007 Paris, France

Blackwell Wissenschafts-Verlag GmbH
  Kurfürstendamm 57
  10707 Berlin, Germany

  Zehetnergasse 6
  A-1140 Wien
  Austria

DISTRIBUTORS

Marston Book Services Ltd
PO Box 269
Abingdon
Oxon OX14 4YN
(*Orders*: Tel: 01235 465500
             Fax: 01235 465555)

USA
Blackwell Science, Inc.
238 Main Street
Cambridge, MA 02142
(*Orders*: Tel: 800 215-1000
              617 876-7000
            Fax: 617 492-5263)

Canada
Copp Clark, Ltd
2775 Matheson Blvd East
Mississauga, Ontario
Canada, L4W 4P7
(*Orders*: Tel: 800 263-4374
              905 238-6074)

Australia
Blackwell Science Pty Ltd
54 University Street
Carlton, Victoria 3053
(*Orders*: Tel: 03 9347-0300
            Fax: 03 9347-3016)

First published 1996

Set in 10 on 13pt Times
by Aarontype Ltd, Bristol
Printed and bound in Great Britain
by Hartnolls Ltd, Bodmin, Cornwall

The Blackwell Science Logo is a trade mark of
Blackwell Science Ltd, registered at the United
Kingdom Trade Marks Registry

A catalogue record for this title
is available from the British Library

ISBN 0-632-04088-2

Library of Congress
Cataloging-in-Publication Data

Davies, J.M.
  Plastic design to BS 5950/the Steel Construction
  Institute: by
J.M. Davies and B.A. Brown.
     p.   cm.
  Includes bibliographical references.
  ISBN 0-632-04088-2
  1. Building, Iron and steel – Standards.
  2. Structural design – Standards.
  3. Plastic analysis (Engineering) I. Brown, B.A.
  (Brian A.) II. Steel Construction Institute
  (Great Britain)
III. Title.
TA684.D36 1996
624.1'821–dc20   96-8475
                                            CIP

Contents

Foreword

It is estimated that 50% of the constructional steelwork used in the UK is fabricated into single-storey buildings. Within this major market sector the portal frame is the most common structural form and plastic analysis is the most frequently used method for its design. Plastic design is also widely used for continuous beams and other simple indeterminate structures.

The plastic design of structural steelwork has come a long way since it was first allowed in the UK by the 1949 edition of BS 449. Yet it was only with the publication of BS 5950 in 1985, nearly 40 years later, that designers in the UK were given code of practice clauses with which to put the design on a more formal basis. There were considerable benefits from this state of affairs because it allowed the practical usage of plastic theory to develop without the constraints of a rigorous design code. There have also been disadvantages in that some of the procedures that have evolved into practice have been more in the nature of folklore than sound engineering.

BS 5950 introduced a higher yield stress for the commonly used grades of mild steel together with lower load factors. These, taking advantage also of the relatively new reduction in the snow load, have led to much lighter steel frames than those which were used in earlier practice. This situation requires the designer to pay more attention to member and frame stability as well as to working load deflections. These are considerations that are carefully treated by the authors.

It is unreasonable to expect an initial statement of detailed design procedures to be without flaw or ambiguity and it has to be admitted that the 1985 Edition of BS 5950 was far from perfect. Many of the problem areas have been improved with the 1990 Edition but some inconsistency and lack of clarity still remains and the writers have attempted to make their own contribution to the debate on these where this has been appropriate. In doing this they have been able to draw on two quite different lifetime experiences in the subject.

Professor Davies was introduced to the finer points of plastic theory in the early 1960s when he was a research student studying various aspects of the subject under the supervision of Professor M.R. Horne, one of the early pioneers of plastic theory whose contributions have had a profound influence on the state-of-the-art. Since then, Professor

Davies has retained an active interest in the development of plastic theory. His own significant contributions have included writing the central elastic–plastic analysis and plastic design modules of the market-leading computer package FASTRAK 5950 and some recent important improvements in the understanding of the significance of second-order effects in plastically designed portal frames.

He has also taught the subject to generations of students and is well aware that many graduate with a relatively flimsy knowledge of some of the important basic principles without which plastic design cannot be confidently used. An important aspect of the book, therefore, is a complete development of plastic theory from the elementary principles to a practical tool for the design office.

In contrast, Brian Brown has built up his experience in the design office. With the Conder group in the early days of the subject, he was deeply involved in developing plastic design as a practical tool for the competitive design of portal frames. The Conder companies were widely regarded as leading the market in this respect and, rising to the office of Technical Director, Brian Brown has built up a unique experience of using plastic design in practice and has made numerous contributions to the way things are now done.

Combining together, Davies and Brown have been able to write a book with a unique blend of theory and practice which will be of equal benefit to students and practitioners alike. It combines a complete university course to Master level in the subject with all of the practical methods that designers will require. Emphasis is entirely on manual methods – designers do not need to know how to write a software package. If they have a computer, they do need to know how to do quick manual checks on the results that it produces.

There are a number of aspects of this that are unique and two are worthy of special mention. It recognises the increasing significance of the elastic critical load as a measure of the sensitivity of a structure to second-order effects. This is implicit in some of the sections of BS 5950 dealing with plastic theory. It is much more explicit in Eurocode 3 and designers will clearly have to become more confident in using this parameter in their design procedures. The importance of this can be measured by the fact that some structures that have been designed and built on the basis of plastic theory in recent years cannot be justified when proper consideration is given to second-order effects. Such structures probably only survive as a consequence of the stiffness of their cladding! Secondly, while much of the book is devoted to portal frames, Chapter 6 contains the first attempt to give a complete plastic design procedure for multi-storey frames.

This book is therefore a very worthwhile reference for all those who wish to have a detailed understanding of practical plastic design. It should be of lasting benefit to practising engineers, students and their teachers. I commend it to you warmly.

Dr Graham Owens
Director
The Steel Construction Institute

Chapter 1

Introduction

1.1 Elastic and plastic properties of structural steel

Plastic design takes advantage of an important and unique property of mild steel, namely its ductility. Figure 1.1(a) shows a typical stress–strain curve obtained from a simple tensile test.

Unless a tensile test is carefully carried out using sophisticated equipment, the upper yield stress is difficult to determine and it has little practical significance. It is therefore ignored in design calculations. Similarly, except for one particular exception which will be considered in Chapter 3, it is conservative and convenient to ignore strain hardening and to consider the simplified stress–strain curve shown in Fig. 1.1(b).

The significant feature in both Figs 1.1(a) and (b) is the long yield plateau which allows the possibility of considerable plastic strain at constant stress. Figure 1.1(a) is typical of the carbon steels that have traditionally been used in plastic design. However, not all structural steels have such a favourable stress–strain curve and Fig. 1.2[1.1] shows the initial part of the stress–strain curve for a range of steels including some of the higher grades. It can be seen that the latter have a more gradually yielding behaviour though still exhibiting considerable ductility. Provided that adequate ductility is available (see below), such steels can still be used for plastic design with the yield stress replaced by the 0.2% proof stress in the conventional way.

BS 5950: Part 1: 1990 includes three standard grades of steel designated design grades 43, 50 and 55. These steels are defined in the European Standard BS EN 10025: 1993 and their significant properties are given in Table 1.1.

The other properties required for design are as follows:

Modulus of elasticity	$E = 205\,\text{kN/mm}^2$
Poisson's ratio	$\nu = 0.30$
Coefficient of linear thermal expansion	$\alpha = 12 \times 10^{-6}$ per °C
Density	$\rho = 7850\,\text{kg/m}^3$

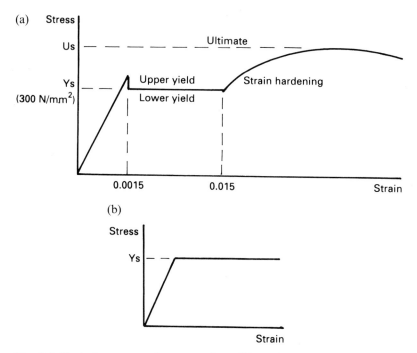

Fig. 1.1 Typical stress–strain curves for mild steel: (a) typical stress–strain curve; (b) simplified stress–strain curve.

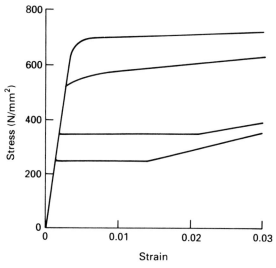

Fig. 1.2 Typical initial stress–strain curves for structural steels[1.1].

Table 1.1 Properties of structural steel grades.

Design grade	Thickness less than or equal to (mm)	Design strength p_y (N/mm^2)	Minimum yield stress Y_s (N/mm^2)	Minimum ultimate strength U_s (N/mm^2)	Minimum elongation on gauge length of $5.65\sqrt{S_0}$
43	16	275	275	410	20–22%
	40	265	265	410	
	63	255	255	410	
	100	245	245	410	
50	16	355	355	490	20–22%
	40	345	345	490	
	63	335	335	490	
	100	325	325	490	
55	16	450	450	550	19%
	25	430	430	550	
	40	415	430	550	
	63	400	415	550	

Notes: 1. In BS 5950, the design strength p_y is taken as equal to the yield stress Y_s but not greater than $0.84\,U_s$.
2. The sub-grades A, B, C, refer to the resistance to brittle fracture as defined by Charpy impact tests.
3. S_0 is the original cross-sectional area of the gauge length.
4. BS 5950: Part 2: 1992 specifies that all hot-rolled structural steel products shall comply with either BS 4360 or BS EN 10025 (grades Fe 360, Fe 430 or Fe 510 only) or with a number of performance requirements.
5. BS EN 10025 uses different terminology to BS 4360 to define the steel grades in terms of strength and other properties.

Steel grades other than those described above may be used for plastic design to BS 5950: Part 1 provided that the following requirements are satisfied:

(a)   the stress–strain diagram has a plateau at the yield stress extending for at least six times the yield strain;
(b)   the ratio of the specified minimum ultimate tensile strength to the specified minimum yield strength is not less than 1.2;
(c)   the elongation on a gauge length of $5.65\sqrt{S_0}$ is not less than 15% where S_0 is the original cross-sectional area of the gauge length.

1.2   Basis of plastic design

Plastic design is applicable to steel structures which carry load predominantly as a consequence of the resistance of their members to bending and is concerned with providing an adequate load factor against collapse of the structures. The basic principles may be illustrated by considering the behaviour of the simply supported beam shown in Fig. 1.3 as the load is increased up to the level where the beam fails.

If w represents the unfactored load per unit length and λ the load factor, the relationship between λ and the deflection under the load is shown in Fig. 1.4. The stress distributions at the centre of the beam at the various stages of loading are shown in Fig. 1.5. Figure 1.5(a) shows a typical cross-section which is assumed to have both horizontal and vertical axes of symmetry. At load levels below the yield load factor λ_y, the stress distribution is linear with a maximum value f in the outermost fibres as shown in Fig. 1.5(b). As the load level is increased above λ_y, the outer fibres yield and yield zones start to spread inwards as shown in Fig. 1.5(c). The load deflection curve is linear elastic up to λ_y but above

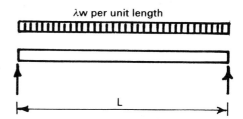

Fig. 1.3 Simply supported beam.

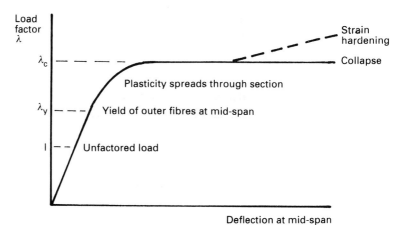

Fig. 1.4 Load-deflection curve for simply supported beam.

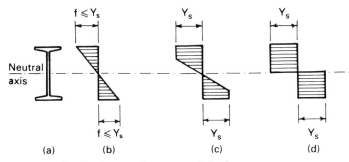

Fig. 1.5 Stress distributions at the centre of the beam.

this value it starts to become non-linear as shown in Fig. 1.4. The inward spread of plasticity continues until the whole cross-section at mid-span has become plastic as shown in Fig. 1.5(d). At this stage, the beam can accept no more load and the load-deflection curve has become horizontal at the collapse value of the load factor λ_c.

It may be noted that if account is taken of the influence of strain hardening, the load-deflection curve may continue to show a small rise at large deflection as shown by the broken line in Fig. 1.4. However, this small beneficial effect is usually ignored. At collapse the region of the beam below the load behaves like a *hinge* with the result that the beam has effectively become a *mechanism* as shown in Fig. 1.6.

This simple example illustrates the two basic principles of plastic theory:

❑ In plastic design, structures are assumed to collapse by the formation of a *collapse mechanism*.
❑ Collapse mechanisms are brought about by the formation of one or more *plastic hinges*. At a plastic hinge, the cross-section has become fully plastic as shown in Fig.1.5(d) with the result that it can rotate at constant bending moment. The bending moment at a plastic hinge is termed the *fully plastic moment of resistance* or, more briefly, the *full plastic moment*.

Consider now the same beam when its ends are built-in to rigid supports as shown in Fig. 1.7. As the load is increased through the elastic

Fig. 1.6 Collapse mechanism for simply supported beam.

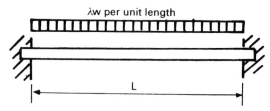

Fig. 1.7 Beam with fixed ends.

range, first yield will occur at the positions of maximum elastic bending moment adjacent to the rigid supports. As the load is increased further, plastic hinges will form at these positions but the centre of the span remains elastic and the beam cannot collapse. This state of affairs is shown in Fig. 1.8(a). As more load is applied, the bending moments at the supports remain constant at their fully plastic values but the moment at mid-span increases. Collapse finally occurs when the bending moment at mid-span reaches its fully plastic value as shown in Fig. 1.8(b). The load-deflection curve for this beam is shown in Fig. 1.9.

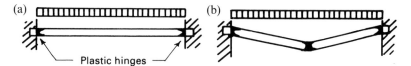

Fig. 1.8 Spread of plasticity in beam with fixed ends: (a) centre elastic, ends plastic; (b) collapse mechanism with three plastic hinges.

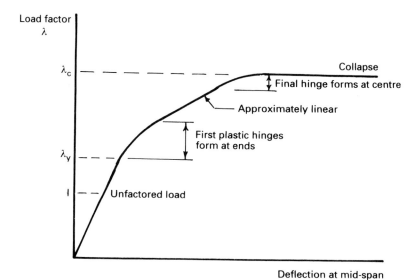

Fig. 1.9 Load-deflection curve for beam with fixed ends.

Initially the curve is linear elastic but becomes non-linear as plasticity spreads through the section at the fixed ends. When the plastic hinges at the ends are fully formed, the curve becomes approximately linear again at a reduced slope as these hinges rotate and bending moments are redistributed towards the centre of the beam. When the third plastic hinge has formed at mid-span, a collapse mechanism exists and the load-deflection curve is horizontal. The structure can continue to deform at constant load.

Much of the theory which follows will be concerned with the prediction of collapse mechanisms and the determination of the corresponding load factors at collapse. Figure 1.10 shows some typical collapse mechanisms.

It is important to note that, in each of the collapse mechanisms shown in Fig. 1.10, sufficient plastic hinges have formed to transform at least part of the structure into a valid mechanism capable of deflecting indefinitely without any change in the loads. It is also important to note that we generally proceed direct to the collapse condition without considering the intermediate elastic-plastic conditions prior to collapse. Thus, plastic design provides consistency of the load factor against collapse whereas the ultimate strength of an elastically designed structure is variable depending on the elastic bending moment distribution.

It is implicit in the above discussion that the structure is sufficiently robust for it to achieve a mechanism form of collapse without the adverse influence of any form of *instability*. Three separate forms of

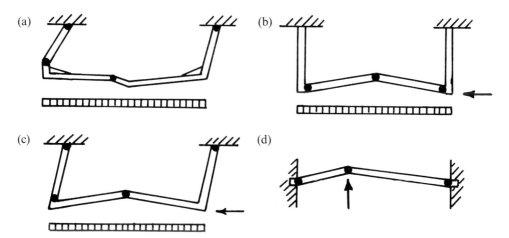

Fig. 1.10 Typical collapse mechanisms: (a) built-in beam with off-centre point load; (b) simple portal frame; (c) alternative mechanism for portal frame (b); (d) pitched roof portal frame with uniformly distributed load.

instability are possible in steel structures and each must be guarded against. Their particular characteristics are as follows:

❑ *Local instability* may arise if the elements of the cross-section (flanges or web) are too slender for plastic hinges to form and rotate at constant bending moment without the occurrence of local buckling at the plastic hinge positions. Sections which are sufficiently compact for plastic hinge action are termed 'plastic sections' and simple rules defining plastic sections are given in Table 7 of BS 5950 Part 1.

For the typical shapes shown in Fig. 1.11, the requirements for a plastic section may be summarised as follows:

For rolled sections $\qquad \dfrac{b}{T} \leq 8.5\epsilon$

For sections built up by welding $\qquad \dfrac{b}{T} \leq 7.5\epsilon$

For webs generally $\qquad \dfrac{d}{t} \leq \dfrac{79\epsilon}{0.4 + 0.6\alpha}$

where $\epsilon = \sqrt{275/p_y}$, $p_y = $ design strength of steel $(=Y_s)$, and $\alpha = 2y_c/d$.

❑ *Lateral instability* may arise in members which are insufficiently braced laterally so that they are relatively slender with regard to buckling about their minor axis. In plastic design, it is necessary to ensure that lateral instability is prevented by the choice of suitable sections and by the presence of adequate bracing. A particular requirement that should be noted in this connection is (clause 5.3.5) that torsional restraints should be provided at all plastic hinge locations or, if this is impracticable, within a distance of half the section depth of the plastic hinge location.

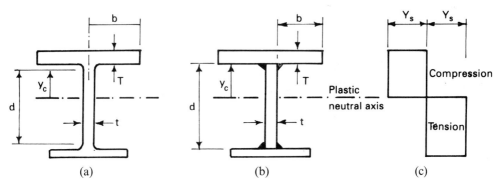

Fig. 1.11 Typical plastic sections: (a) rolled section; (b) section built up by welding; (c) stress distribution.

❑ *Frame instability* occurs in the plane of the frame as a consequence of significant axial compressive loads together with finite deflections. In many structures designed by plastic theory, the axial compressive loads are relatively small and this effect can be discounted. In other structures, it is necessary to predict the reduction in the collapse load occasioned by frame instability (also termed second-order effects) and there are alternative ways of dealing with this which will be mentioned at appropriate places in the book.

1.3   Evaluation of the full plastic moment

In order to demonstrate the principles involved in the determination of the full plastic moment, M_p, a general section with a vertical axis of symmetry Y–Y will be considered as shown in Fig. 1.12(a). The section will be assumed to be subject to a pure bending moment with no axial load. Consider the spread of plasticity as the moment is increased to its full plastic value.

Figure 1.12(b) shows the stress distribution in the elastic range of loading according to the engineer's theory of bending. The distribution is linear, having a zero value at the neutral axis and a maximum value at one or other of the outer fibres. As the bending moment increases, yield takes place in the region of maximum stress and the yielded zone starts to spread inwards until the yield stress is reached at the other extremity. Figure 1.12(c) shows the stress distribution when there is a yielded zone in the lower part of the section and the yield stress has just been attained in the uppermost fibres.

Further increases in bending moment result in the yielded zones spreading inwards until they meet, as shown in Fig. 1.12(d). At this

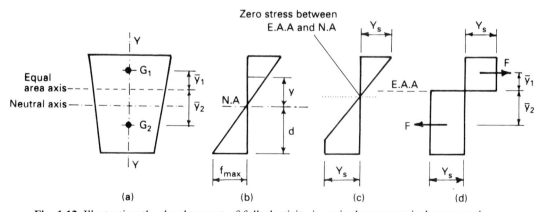

Fig. 1.12 Illustrating the development of full plasticity in a singly symmetrical cross-section.

stage the section is fully plastic and the bending moment is equal to the fully plastic moment of resistance M_p. The stress distribution consists of a compressive stress block in the upper part of the cross-section and a tensile stress block in the lower part of the section, both blocks being at the yield stress Y_s. As the section is subject to pure bending with zero axial force, the forces F associated with each stress block must be equal and, furthermore, if the cross-sectional area is A,

$$F = Y_s \frac{A}{2} \tag{1.1}$$

It also follows that, in Fig.1.12(d), the zero stress axis divides the cross-section into two equal areas so that it is usually termed the 'equal area axis' (EAA). As plasticity has spread through the section the zero stress axis has moved from the original neutral axis passing through the centroid of the section to the equal area axis. If $\bar{y}_1$ and $\bar{y}_2$ are distances from this axis to the centroids of the individual stress blocks, it follows that the full plastic moment M_p is given by

$$M_p = Y_s \frac{A}{2} (\bar{y}_1 + \bar{y}_2) \tag{1.2}$$

Computationally, it is more convenient to consider the section to be divided into a number of elementary areas A_i each having a centroid distant y_i from the equal area axis. It then follows that

$$M_p = Y_s \sum_i A_i y_i \tag{1.3}$$

that is

$$M_p = Y_s S_x \tag{1.4}$$

where S_x = first moment of area of the cross-section which is a section property termed the *plastic section modulus*. The plastic section moduli of all standard hot-rolled sections are tabulated in standard publications such as *Steelwork Design Guide to BS 5950*[1.2] so that for such sections the evaluation of M_p is trivial.

It may be noted that the yield moment M_y at a cross-section is given by

$$M_y = Y_s Z \tag{1.5}$$

where Z = elastic section modulus. It follows that there is a further dimensionless section property termed the *shape factor* α which is defined by

$$\alpha = \frac{S_x}{Z} = \frac{M_p}{M_y} \tag{1.6}$$

For Universal Beam sections bent about the major axis, α generally lies within the range 1.15 to 1.2. For other sections considerably larger values are possible.

Example 1.1   Find the shape factor of the Tee section shown in Fig. 1.13.

(a)   Determination of the elastic section modulus Z.
Let the neutral axis be a distance $\bar{y}$ from the top of the flange.

$$\text{Area} = 12.5(100 + 87.5) = 2344 \, \text{mm}^2$$

$$\therefore \quad 2344 \, \bar{y} = 87.5 \times \frac{12.5^2}{2} + 12.5 \times \frac{100^2}{2}$$

$$\therefore \quad \bar{y} = 29.8 \, \text{mm}$$

$$\text{Second moment of area} = \frac{87.5 \times 12.5^3}{12} + 87.5 \times 12.5 \times 23.33^2$$

$$+ \frac{12.5 \times 100^3}{12} + 12.5 \times 100 \times 20.42^2$$

$$= 2\,172\,400 \, \text{mm}^4$$

$$\therefore \quad Z = \frac{2\,172\,400}{70.42} = 30\,850 \, \text{mm}^3$$

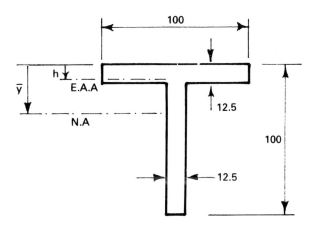

Fig. 1.13 Tee section.

(b)   Determination of the plastic section modulus S_x.
Let the equal area axis be a distance h from the top of the flange (it obviously lies just within the flange). Then

$$100h = 87.5 \times 12.5 + 100(12.5 - h)$$

$$\therefore \quad h = 11.7\,\text{mm}$$

and

$$S_x = \frac{100 \times 11.7^2}{2} + \frac{87.5 \times 0.8^2}{2} + \frac{12.5 \times 88.3^2}{2} = 55\,600\,\text{mm}^3$$

(c)   Determination of shape factor α

$$\alpha = \frac{S_x}{Z} = \frac{55\,600}{30\,850} = 1.80$$

1.4   Reduction of the full plastic moment due to an axial thrust

For sections with only one axis of symmetry, the theory describing the interaction between bending moment and axial force is extremely complicated so that consideration will be restricted to sections that are symmetrical about both horizontal and vertical axes. It is convenient to explain the procedure first for a rectangular cross-section and then to generalise the procedure for, for example, I-sections.

Consider the rectangular cross-section shown in Fig. 1.14(a) which is assumed to be fully plastic under the combination of a compressive force P and a bending moment M'_p which is less than the full plastic moment M_p.

Clearly, the greater part of the cross-section will be in plastic compression, as shown. The stress blocks are shown in Fig. 1.14(b) with the compressive block divided into a shaded portion resisting the moment and an unshaded portion of depth a resisting the axial force P so that

$$P = Y_s ba \tag{1.7}$$

We define the mean axial stress p by

$$P = pbd \tag{1.8}$$

$$\therefore \quad \frac{a}{d} = \frac{p}{Y_s} = n \tag{1.9}$$

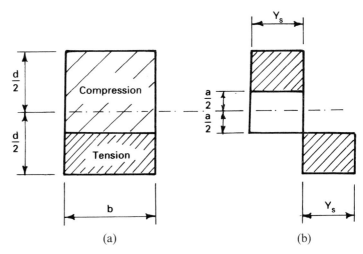

Fig. 1.14 Rectangular section subject to moment and axial force: (a) cross-section; (b) stress blocks.

where n is the ratio of the mean axial stress to the yield stress and varies between 0 (for no axial force) and 1 (section fully plastic in compression). The effective plastic section modulus S' is the first moment of area of the shaded portions.

$$\therefore \quad S' = 2b\left(\frac{d-a}{2}\right)\left(\frac{d+a}{4}\right) = \frac{bd^2}{4}(1-n^2) \tag{1.10}$$

and

$$M'_p = Y_s S' \tag{1.11}$$

In Equation (1.10), the expression $bd^2/4$ is the plastic section modulus of the rectangular cross-section and $(1-n^2)$ is a reduction factor for axial thrust. Evidently, a considerable axial thrust is required before the reduction in M_p becomes significant. This is true for I-sections as well as rectangular sections so that in many low-rise framed structures, where the axial forces are not large, it may be permissible to ignore this effect.

It may be noted that the result given in Equation (1.10) may alternatively be obtained by deducting from the section modulus of the gross section the section modulus of the portion resisting the axial thrust, i.e.

$$S' = \frac{bd^2}{4} - \frac{ba^2}{4} = \frac{bd^2}{4}\left(1 - \frac{a^2}{d^2}\right) \tag{1.12}$$

This alternative technique is useful in dealing with I-sections as shown in the following example.

Example 1.2   Find the expressions for the reduction due to axial thrust for the I-section shown in Fig. 1.15.

(a)   Determination of the unreduced plastic section modulus S_x

$$S_x = \frac{10 \times 140^2}{4} + 2(100 \times 10 \times 75) = 199\,000\,\text{mm}^3$$

(b)   For the case where the plane of zero stress is in the web, let the axial thrust P (N) be carried by a depth a (mm) of the web as shown in Fig. 1.16.

Gross area of cross-section $= 3400\,\text{mm}^2$

$$\therefore \quad P = 3400p = 10aY_s$$

$$\text{i.e.} \quad n = \frac{p}{Y_s} = \frac{a}{340}$$

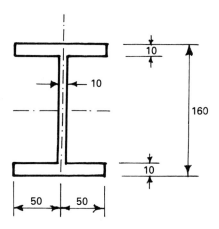

Fig. 1.15 I-section.

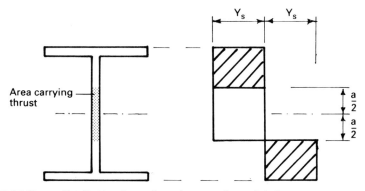

Fig. 1.16 Stress distribution in an I-section carrying axial thrust.

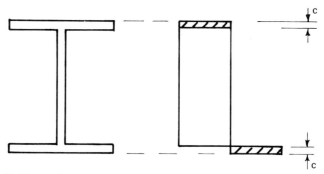

Fig. 1.17 Plane of zero stress in the flanges.

The reduction in section modulus as a result of the axial thrust is

$$\frac{10a^2}{4} = \frac{10 \times 340^2 n^2}{4} = 289\,000n^2 \text{ mm}^3$$

The reduced plastic section modulus S'_x is therefore given by

$$S'_x = 199\,000 - 289\,000n^2 \text{ mm}^3$$

(c) The above expression is valid as long as the plane of zero stress remains in the web, i.e. for $a \le 140$, i.e. for $n \le 140/340 = 0.412$.

(d) When the plane of zero stress is in the flange, as shown in Fig. 1.17

$$P = 3400p = (3400 - 200c)\,Y_s$$

$$\therefore \quad n = \frac{p}{Y_s} = \frac{3400 - 200c}{3400}$$

$$\therefore \quad c = 17(1 - n)$$

The reduced plastic section modulus S'_x is therefore given by

$$S'_x = 100c(160 - c) = 1700(1 - n)(143 + 17n) \text{ mm}^3$$

Obviously, it would be impractical to require calculations of the above nature for all Universal Beams and other I-sections. They are therefore tabulated for all doubly symmetrical rolled sections in precisely the form derived in the above example.

Low axial stresses have a negligible effect on the full plastic moment and can usually be ignored in low-rise structures. In the above example, if the mean axial stress $p = 15\,\text{N/mm}^2$ and the yield strength $p_y = 275\,\text{N/mm}^2$

$$n = 0.0545$$

$$\therefore \quad S'_x = 199\,000 - 289\,000 n^2$$

$$= 198\,140\,\text{mm}^3$$

i.e., $S'_x = 0.995 S_x$

The effect of this quite typical axial stress on the full plastic moment, and therefore on the load factor, is of the order of one half of one per cent.

1.5   Effect of shear

The influence of shear on the full plastic moment of the cross-section can generally be neglected. BS 5950: Part 1 only requires a reduction when the shear force F_v is greater than 0.6 times the shear capacity of the cross-section. This will only be the case in highly loaded beams of relatively short span and relatively large depth. Clause 4.2.6 gives simple design expressions for such cases.

1.6   Unsymmetrical sections

BS 5950: Part 1, clause 5.3.4 specifically requires that where plastic hinges occur in a member, the cross-section should be symmetrical about its axis perpendicular to the axis of hinge rotation.

1.7   Effect of holes

In general, holes such as those required for bolts do not influence the design of members except where they coincide with the position of a plastic hinge (see BS 5950: Part 1, clause 3.3.3). Where they occur within a length equal to the member depth on either side of a plastic hinge location, the full plastic moment should be reduced accordingly. Furthermore, holes in the tension flange near hinge positions, or where local yield lines are assumed in the design of rigid connections, should either be drilled or else punched 2 mm undersize and reamed.

1.8   Loads and load factors

Clause 5.3.2 of BS 5950 Part 1 states that plastic design may be used where the loading is predominantly static so that fatigue is not a design criterion. Otherwise, the loads and load factors are precisely the same as for any other design in structural steelwork as given in Table 2 of BS 5950 Part 1 and repeated, in part, in Table 1.2.

Table 1.2 Load factors.

Loading	Load factor γ_f
Dead load	1.4
Dead load acting with wind and imposed loads combined	1.2
Imposed load	1.6
Imposed load acting with wind load	1.2
Wind load	1.4

Attention may be drawn at this point to the fact that BS 5950 is a limit state code but in a restricted sense, as described in clause 2.1.1. Strictly speaking, the design check at the ultimate limit state should include the following factors:

- ❏ material strength γ_m
- ❏ loading γ_ℓ
- ❏ structural performance γ_p

In BS 5950, the design strength of steel, p_y, is taken to be equal to the yield stress Y_s so that $\gamma_m = 1.0$. Furthermore, γ_ℓ and γ_p are combined into the single load factor γ_f thus simplifying considerably the formal limit state concept.

1.8.1   Snow loads

Snow loads are imposed loads and are specified in BS 6399: Part 3: 1988. They are:

(1)   A uniformly distributed snow load over the complete roof. The intensity of this load depends on the location of the building and the roof slope. A value of $0.6\,\text{kN/m}^2$ applies in many cases but it can be greater than this.

(2) Asymmetrically distributed snow caused by the wind transferring snow from one side of the ridge to the other. This need only be considered when the roof slope exceeds $15°$ and is also $\geq 0.6\,kN/m^2$.

(3) Exceptional snow, which represents infrequent drifting and depends on the roof profile and any abrupt changes in height. Typical situations where exceptional snow must be considered are perimeter parapets, the valleys of multi-span portal frames, and changes in eaves height extending across the width of the building or down the length of the building. This localised snow load frequently exceeds $2.0\,kN/m^2$ and is more likely to affect the design of the purlins and sheeting than the design of the members of the main frame. BS 6399 recommends the use of a reduced load factor for exceptional snow and it is accepted practice to use a value of 1.05 which is specified for other exceptional loadings in clause 2.4.5.4 of BS 5950. Exceptional snow is only considered at the ultimate limit state where it is combined with 1.4 times the dead load.

1.8.2 Notional horizontal loads

'Notional horizontal loads' are defined in clause 2.4.2.3 of BS 5950 and these should be applied to all structures in order to represent imperfections such as lack of verticality of the stanchions. They are equivalent to about double the usual erection tolerance. The horizontal loads are normally 0.5% of the factored vertical load and should be applied at each roof and floor level or their equivalent. Notional horizontal loads have been the cause of some confusion, mainly because of some unfortunate wording in the earlier (1985) edition of BS 5950. In the 1990 edition, the crucial clauses 2.4.2.3 and 5.1.2.3 include important changes which have clarified the position.

It is now mandatory that the notional horizontal loads should be considered together with the full factored dead and live loads when considering the vertical load case and this is the critical load case in most pitched roof portal structures. However, these loads are quite small and generally have little effect on the design of low-rise portal frames. Possibly for this reason or perhaps in ignorance, most designers of portal frames do not include these loads when, according to BS 5950, they should. In any event, the notional horizontal loads are more significant in multi-storey construction.

Further confusion may arise until it is realised that the same notional horizontal forces are used conveniently and quite separately in sway stability checks in clauses 5.1.3 and 5.5.3.2.

1.8.3   Crane loads

The load factors for crane loads are given in Table 2 of BS 5950: Part 1 and some guidance on their application is given in clause 2.4.1.2. In general, a structure with overhead travelling cranes is likely to be designed for both crane loads and imposed loads and this gives rise to a large number of possible load combinations. Reference 1.3 lists no less than 15 of these and notes that there may be even more if the crane loads and imposed loads are not additive. The most important combinations are:

(1) $1.4D + 1.6I$
(2) $1.4D + 1.4W$
(3) $1.4D + 1.6V$
(4) $1.4D + 1.6H$
(5) $1.4D + 1.4V + 1.4H$
(6) $1.2D + 1.6I + 1.6V$
(7) $1.2D + 1.6I + 1.6H$
(8) $1.2D + 1.6I + 1.4V + 1.4H$
(9) $1.2D + 1.2I + 1.2W + 1.2V + 1.2H$

where D = dead load, W = wind load, V = vertical crane load (static wheel loads plus allowance for dynamic effects), and H = horizontal crane load (three alternatives).

Evidently, especially when using manual design methods, the engineer should use some judgement in order to avoid processing obviously non-critical load combinations.

Note:[1.4]   It is *not* correct to separate the crane wheel loads into components due to the lifted load and the dead weight of the crane, applying a load factor of 1.6 only to the lifted load and 1.4 to the dead weight. The factor of 1.6 for the vertical crane load in Table 2 should be applied to the total wheel loads from the crane.

1.9   Deflections

Strictly speaking, deflection is only one of a number of factors which should be checked at the serviceability limit state. However, in practice, deflection is the only serviceability limit that is significant in the vast majority of structures which may be designed using plastic theory.

Generally, the serviceability loads are the unfactored imposed loads. However, when considering the case of dead plus imposed plus wind load, only 80% of the imposed and wind loads need be considered.

BS 5950 Part 1, clause 2.5.1, includes the following statement:

'The deflection under serviceability loads of a building or part should not impair the strength or efficiency of the structure or its components or cause damage to the finishings.'

It then proceeds (in Table 5) to give recommended deflection limits for certain structural members. However, pitched roof portal frames are specifically excluded and no guidance is given for these particular structures. The reason for this is presumably because, in a low-rise portal frame clad with profiled sheeting, there may be no deflection requirements. The structure and its cladding are able to accept any reasonable deflections that are compatible with the stability requirements. However, the designer can never ignore deflections completely and it would be very unwise to design any portal frame structure without an estimate of the likely deflections. Some situations where deflections may influence the design are:

❑ crane buildings;
❑ buildings with masonry walls;
❑ in large span buildings, the relative vertical deflection between the gable and the first frame may be excessive for light gauge steel purlins;
❑ in tall buildings, the relative sway between the gable and the first frame may set up unacceptable stressed skin forces in the cladding and its fasteners.

It should also be realised that, if a structure is too flexible, second-order effects become significant. Portal frames are not exempt from this general rule and, particularly in frames with slender rafters, the reduction in the plastic collapse load cannot be ignored. This important point will be considered in more detail in Section 3.2.

Additional guidance on deflections is given in References 1.5 and 1.6. This is subject to continuing discussion within the industry and may, in time, be subject to revision.

1.10   Economy gained by using BS 5950

There are several reasons why the use of BS 5950, and the other related British Standards mentioned in this chapter, result in more economical design than was obtained previously with BS 449. These show to particular effect in the type of structure considered in this book. Thus, BS 5950 has introduced lower load factors, namely 1.4 for dead load and 1.6 for live load, compared with the global factor of 1.7 used previously. Furthermore, as a consequence of changes in BS 4360, the yield stress of mild steel has increased so that the design strength of Grade 43 steel has moved up from 250 N/mm² to 275 N/mm². For a structure in which the imposed load is three times the dead load, these two changes alone result in a reduction in required bending strength of

$$\frac{0.25 \times 1.4 + 0.75 \times 1.6}{1.7} \left(\frac{250}{275}\right) = 0.83$$

In many low-rise structures, further savings arise because of the reduction in the intensity of uniformly distributed snow from its former value of 0.75 kN/m² to the value of 0.6 kN/m² discussed in section 1.8.1. It follows that designs are becoming ever more slender and designers need to pay more careful attention to such factors as frame stiffness, frame stability and member stability when designing to the new codes.

References

1.1 Brockenbrough R L. Material properties, in *Constructional Steel Design – An International Guide*, Eds P J Dowling *et al.* Elsevier Applied Science, 1992, Chapter 1.2.

1.2 Steel Construction Institute. *Steelwork Design Guide to BS 5950: Part 1*: 1990, Volume 1 – Section properties and member capacities, 3rd ed. 1992.

1.3 Advisory Desk AD 111. Load factors for combinations involving crane loading. *Steel Construction Today*, May 1992, pp. 134–135.

1.4 Advisory Desk AD 121. Load factors for crane loads. *Steel Construction Today*, September 1992, p. 225.

1.5 Steel Construction Institute. *Steelwork Design Guide to BS 5950*: Volume 4 – Essential Data for Designers, 1991.

1.6 Advisory Desk AD 090. Deflection limits for pitched roof portal frames. *Steel Construction Today*, July 1991, pp. 203–206.

Chapter 2
Principles of Plastic Design

2.1 Criteria for a valid collapse mechanism

In Section 1.2, it has been shown that the basis of plastic design is that structures are assumed to collapse by the formation of sufficient plastic hinges to create a collapse mechanism. The fundamental problems of plastic design are therefore:

❑ prediction of the correct collapse mechanism;
❑ determination of the load factor at collapse;
❑ determination of the bending moment diagram at collapse.

In this chapter, the basic methods of solving these problems will be described. Common to all methods is the fundamental requirement that, at collapse, the following three conditions must be satisfied:

(1) *Equilibrium condition*: The bending moments must represent a state of equilibrium between the internal forces in the structure and the applied loads.
(2) *Mechanism condition:* At collapse, the bending moment must be equal to the full plastic moment of resistance of the cross-section at a sufficient number of sections of the structure for the associated plastic hinges to constitute a mechanism involving the whole structure or some part of it.
(3) *Yield condition:* At every cross-section of the structure, the bending moment must be less than, or equal to, the full plastic moment of resistance.

These conditions are obvious and largely self-explanatory. They correspond to the conditions of *equilibrium* and *compatibility* in the elastic analysis of statically indeterminate structures. It is one of the great advantages of plastic theory that it is not necessary to consider continuity and this generally makes the plastic analysis considerably easier than the elastic analysis of any given structure.

In many structures, there are a number of alternative collapse mechanisms and the correct mechanism is not immediately obvious. It is therefore necessary to approach the correct solution in a series of steps. The following theorems assist the analyst to work towards an acceptable solution. They will merely be stated and illustrated as the proofs are complicated and outside the scope of a practical design manual.

(a) Kinematic Theorem or Minimum Principle

'In the analysis of a structure, an arbitrary choice of collapse mechanism will lead to an estimate of the collapse load which is greater than or equal to the correct one.'

In other words, if we do not know the correct collapse mechanism and make a guess, the solution obtained only represents an upper bound on the collapse load factor λ_c and is potentially *unsafe*. Methods based on assumed collapse mechanisms generally satisfy only the equilibrium and mechanism conditions.

(b) Static Theorem or Maximum Principle

'An arbitrary equilibrium condition which also satisfies the yield condition will lead to an estimate of the collapse load which is less than or equal to the correct one.'

In other words, satisfying the equilibrium and yield conditions without necessarily obtaining a mechanism is essentially a *safe* procedure.

(c) Uniqueness Theorem

'The value of the collapse load which satisfies the three conditions of equilibrium, mechanism and yield is unique.'

It is impossible to obtain for any other load a bending moment diagram which also satisfies these three conditions.

The following simple example serves to illustrate the above theorems.

Example 2.1 If the full plastic moment M_p is 78.0 kNm, find the load factor against collapse for the beam shown in Fig. 2.1. There are two distinct approaches to this problem.

Method 1 combines free and reactant bending moment diagrams to obtain solutions that satisfy the equilibrium and yield conditions as shown in Fig. 2.2.

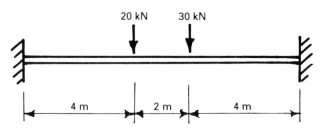

Fig. 2.1 Beam.

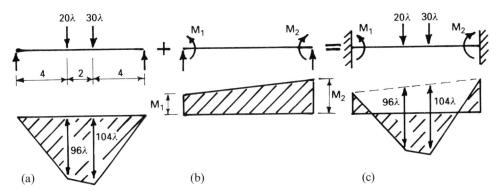

Fig. 2.2 Solution using free and reactant bending moment diagrams: (a) Free bending moments + (b) Reactant bending moments = (c) Actual moments.

In this case it is trivial to also satisfy the mechanism condition with plastic hinges at the ends of the beam where $M_1 = M_2 = M_p$ and with a third plastic hinge under the larger load. It follows that $104\lambda = 2M_p = 156\,\text{kNm}$, i.e. $\lambda = 1.5$, and that the bending moment diagram at collapse and the collapse mechanism are as shown in Fig. 2.3.

Method 2 considers all possible collapse mechanisms and chooses the one with the smallest load factor. Equilibrium between bending

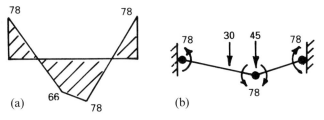

Fig. 2.3 Collapse of single span beam: (a) Bending moments at collapse; (b) Collapse mechanism.

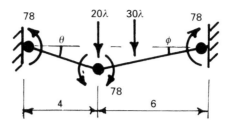

Fig. 2.4 Mechanism (a).

moments and applied loads is satisfied by using the virtual work method shown in Fig. 2.4.

Here the correct mechanism is obvious but, for illustrative purposes, we also consider the improbable alternative. The method considers a virtual displacement of the mechanism that is under investigation assuming that all internal strain is concentrated in the plastic hinges. The members between the plastic hinges are assumed to be perfectly rigid and to make no contribution to the internal work. Thus, for the improbable mechanism:

For compatibility,

$$4\theta = 6\phi$$

By virtual work, external loads × corresponding displacements = hinge moments × rotations:

$$20\lambda \times 4\theta + 30\lambda \times 4\phi = 78(2\theta + 2\phi)$$

i.e. $\quad 120\lambda\phi + 120\lambda\phi = 78 \times 5\phi$

i.e. $\quad \lambda = \dfrac{390}{240} = 1.625$

Similarly, for the other possible collapse mechanism shown in Fig. 2.5:

For compatibility

$$4\theta = 6\phi$$

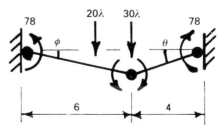

Fig. 2.5 Mechanism (b).

∴ by virtual work

$$30\lambda \times 4\theta + 20\lambda \times 4\phi = 78(2\theta + 2\phi)$$

i.e. $180\lambda\phi + 80\lambda\phi = 78 \times 5\phi$

i.e. $\lambda = \dfrac{390}{260} = 1.5$

We have now considered all possible collapse mechanisms and, by appealing to the minimum principle, can confidently choose the mechanism with the lowest load factor giving, as before, $\lambda = 1.5$.

It should be noted that no design calculation is complete until the bending moment diagram at collapse has been drawn. For this problem, once the load factor and collapse mechanism are known, this is elementary and the resulting bending moment diagram has already been given with method 1.

2.2 Plastic analysis of continuous beams

It has already been stated that one of the advantages of plastic theory is that there are no compatibility conditions in the analysis. As plastic hinges form, they destroy the continuity of the deflection profile and an important consequence of this in continuous beams is that *the collapse of any span is independent of the adjacent spans*.

Consider the important case of a uniform beam that is continuous over several equal spans as shown in Fig. 2.6. As before, w is the unfactored load which is multiplied by a load factor λ and it is required to find the value of λ at collapse.

There are only two collapse cases to consider as shown in Fig. 2.7. Each of the internal spans has exactly the same collapse mechanism and collapse load as shown in Fig. 2.7(a). One end span is the mirror-image of the other as shown in Fig. 2.7(b).

Load λw per unit length

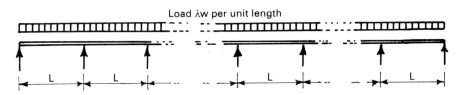

Fig. 2.6 Uniformly loaded continuous beam.

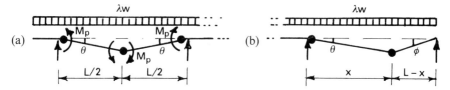

Fig. 2.7 Alternative collapse cases for continuous beams: (a) collapse of interior span; (b) collapse of end span.

For the interior span, using the work equation method (although the method using free and reactant bending moment diagrams is equally simple),

$$\lambda w L \frac{L\theta}{4} = M_p(\theta + 2\theta + \theta)$$

i.e. $$M_p = \frac{\lambda w L^2}{16}$$

if it is a design calculation that is required

or $$\lambda = \frac{16 M_p}{w L^2}$$

if an analysis of an existing design is required.

The left-hand side of the work equation, representing the virtual work done by the uniformly distributed load, sometimes causes problems for those unfamiliar with the method. It is important to remember that we are considering a virtual mechanism in which the members remain perfectly straight between plastic hinges. The distributed load can therefore be considered to be concentrated at the centroids of the straight lengths of member as shown in Fig. 2.8. The derivation of the external work term is then obvious.

The collapse conditions in the end span are not so obvious because it is not clear where the internal plastic hinge is located. The kinematic theorem implies that we must consider all possible positions of this

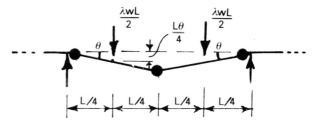

Fig. 2.8 Illustrating the derivation of the external work term.

hinge and choose the one which gives the lowest load factor (or the highest value of M_p in a design calculation). This can be conveniently done by choosing an arbitrary hinge position defined by the variable x, as shown in Fig. 2.7(b), and using the methods of calculus.

The rotations θ and ϕ in Fig. 2.7(b) are related by considering the vertical deflection at the hinge position, that is

$$x\theta = (L - x)\phi$$

The work equation can then be written down including the variable x:

$$\lambda wL \frac{x\theta}{2} = M_p(2\theta + \phi) = M_p\left(2 + \frac{x}{L - x}\right)\theta$$

i.e. $M_p = \dfrac{\lambda wL}{2} \dfrac{x(L - x)}{2L - x}$

The critical value of x is that which maximises M_p, i.e.

$$\frac{dM_p}{dx} = 0$$

and the condition for this is

$$(2L - x)(L - 2x) - x(L - x)(-1) = 0$$

$$\text{i.e.} \quad x^2 - 4xL + 2L^2 = 0$$

$$\text{i.e.} \quad x = L(2 \pm \sqrt{2})$$

This gives a unique root within the span $(0 \le x \le L)$ of

$$x = L(2 - \sqrt{2}) = 0.586L$$

Hence, by back-substitution into the above equation for M_p,

$$M_p = \frac{\lambda wL^2}{11.66}$$

or $\lambda = \dfrac{11.66 M_p}{wL^2}$

As the end-bay condition arises regularly in practical design, this is a useful result that should be noted for future reference.

It is immediately clear that if a beam is designed to be fully continuous over several spans, the end bays require a considerably stronger section than the internal bays ($M_p = \lambda wL^2/11.66$ compared with $\lambda wL^2/16$). The fabrication of connections that are adequate to

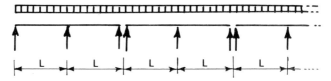

Fig. 2.9 Alternative design to save fabrication costs.

ensure full continuity is expensive and a fully continuous uniform beam may well not be the optimum practical solution. There are several possibilities, including the following:

(1) Design a uniform beam based on the end spans accepting that the internal spans will be over-designed. This may be the best solution if deflections are important and detailing difficulties preclude the use of different sections within the length of the beam.

(2) Design the beam as a series of double spans, as shown in Fig. 2.9. Although this may result in extra costs for the material, these may be more than offset by the savings in fabrication.

Note. It should be appreciated that this arrangement will give rise to reactions which will be alternately $0.75wl$ and $1.25wl$. This may have undesirable consequences in terms of the sizes of the supporting members and foundations.

(3) Use a stronger section for the end span. With this solution, care needs to be exercised with the positioning of the splice. Ideally, the splice should be positioned at the point of contraflecture in the penultimate span, as shown in Fig. 2.10(a). If it is necessary to

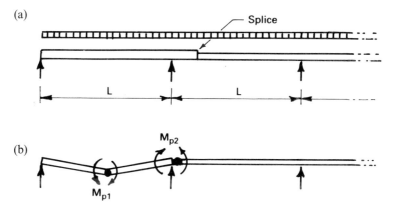

Fig. 2.10 Continuous beam with stronger end span: (a) Splice at point of contraflecture; (b) Illustrating the consequences of a splice at the support.

splice at the penultimate support, it must be appreciated that the plastic hinge at the support will form in the weaker span, as shown in Fig. 2.10(b). The design calculations must be amended to take account of this.

(4) Adjust the length of the end bay (L_1) relative to the internal bays (L_2). This will occasionally be possible and, when it is, it provides a neat solution. The condition for the end bays to collapse under the same load as the internal bays is

$$\frac{\lambda w L_1^2}{11.66} = \frac{\lambda w L_2^2}{16}$$

i.e. $\dfrac{L_2}{L_1} = \sqrt{\dfrac{16}{11.66}} = 1.17$

For more general cases of unequal spans or more complex loading, the designer has the option of using either the work equation method or using free and reactant bending moment diagrams. In either case, the procedure involves considering one span at a time and it will be illustrated by means of the following example using free and reactant bending moment diagrams.

Example 2.2 The continuous beam ABCD in Fig. 2.11 has a uniform section. If collapse just occurs under the loads shown, determine the value of the full plastic moment.

If real hinges are inserted at the supports, the free bending moment diagram can be drawn (in kNm) as shown in Fig. 2.12. As each span is

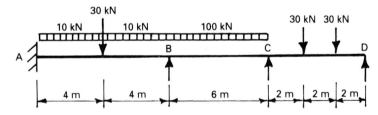

Fig. 2.11 Continuous beam.

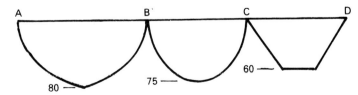

Fig. 2.12 Free bending moment diagram.

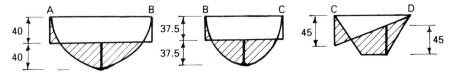

Fig. 2.13 Bending moment diagrams at collapse of individual spans.

simply supported, and therefore statically determinate, this is elementary. The collapse of each span in turn can be considered by drawing a reactant line that is compatible with a mechanism in that span. This will involve a hogging hinge (real or plastic as appropriate) at each support and a sagging plastic hinge somewhere in the span as shown in Fig. 2.13.

The technique of subtracting an inverted reactant diagram from the free diagram should be noted. It is immediately clear that the right-hand span requires the largest full plastic moment (45 kNm) and, by the kinematic theorem, this is the required solution.

It is now possible to draw a whole range of bending moment diagrams for the complete beam which satisfy the uniqueness theorem as shown in Fig. 2.14. The bending moment diagram in the right-hand span is fixed, but in the other two spans, the free and reactant line method automatically satisfies the equilibrium conditions and any reactant line which does not violate the yield condition is admissible. The reactant line for these two spans may therefore lie anywhere in the shaded region.

It should now be possible to appreciate that the collapse conditions in any span are the same regardless of whether there is any settlement of supports or other changes from the assumed 'perfect' structure. This is in contrast to the elastic conditions which can be strongly influenced by movements of supports, lack of rigidity in joints or imperfect fit of component parts and provides one of the merits of plastic analysis. The reason for this is that the formation of plastic hinges systematically releases the compatibility constraints so that each span is statically determinate at collapse.

Note, however, that the settlement of supports or flexibility in the joints will influence the deflections at the serviceability limit. Though not

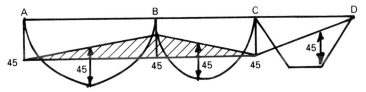

Fig. 2.14 Bending moment diagram(s) at collapse.

directly relevant to 'simple' plastic theory, these factors also influence the plastic hinge history and the elastic-plastic bending moment diagram.

2.3   Simple portal frames

When plastic analysis is extended to portal frames, the collapse mechanism is no longer immediately obvious and it is usually necessary to consider several possibilities. For simple rectangular frames, the work equation method is generally the most advantageous although the method of free and reactant moment diagrams will be reintroduced later when pitched roof portal frames are considered. The considerations will be introduced with reference to a simple example.

Example 2.3   The frame shown in Fig. 2.15 has a uniform full plastic moment of 20 kNm. The loads shown are unfactored, find the load factor λ against collapse.

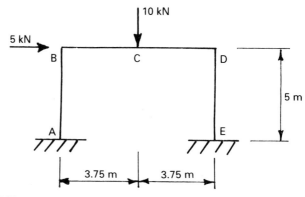

Fig. 2.15 Frame.

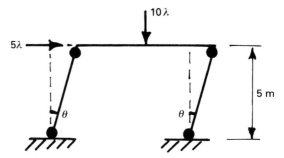

Fig. 2.16 Sway mechanism.

The only possible locations for plastic hinges are at A, B, C, D and E where there is a change of slope in the bending moment diagram. It is impossible for plastic hinges to form between these points which are termed *critical sections*. It then follows that there are only three valid mechanisms which have plastic hinges at the critical sections. These are:

(1)   Sway mechanism (Fig. 2.16)

For a small, rigid-link, movement of the mechanism, the beam moves bodily sideways and there is no downward movement of the 10 kN load. The work equation is therefore

$$5\lambda \times 50 = 20(\theta + \theta + \theta + \theta)$$

i.e. $\lambda = \dfrac{80}{25} = 3.2$

Note.   In plastic collapse mechanisms, the bending moment at a plastic hinge is always related to the direction of rotation of that hinge with the result that *plastic hinges always do positive virtual work*.

(2)   Beam mechanism (Fig. 2.17)

For a *small* movement of the mechanism, the stanchions remain vertical and there is no movement of the 5 kN load. The work equation is therefore

$$10\lambda \times 3.75\theta = 20(\theta + 2\theta + \theta)$$

$$\lambda = \dfrac{80}{37.5} = 2.13$$

(3)   Combined mechanism (Fig. 2.18)

As the name suggests, this is a combination of (1) and (2). It is essential to carry out this com-bination in such a way that the two

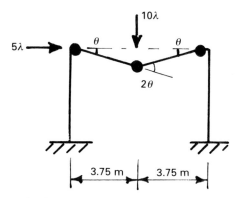

Fig. 2.17 Beam mechanism.

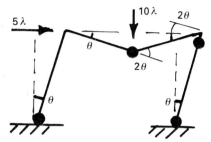

Fig. 2.18 Combined mechanism.

hinges at B are eliminated and replaced by a rigid joint. If this were not the case, the mechanism would have two degrees of freedom and it would be impossible to relate all the movements in the virtual mechanism to a single variable θ. The work equation is therefore

$$5\lambda \times 5\theta + 10\lambda \times 3.75\theta = 20(\theta + 2\theta + 2\theta + \theta)$$

$$\lambda = \frac{120}{62.5} = 1.92$$

As all possible mechanisms have been considered, it follows from the minimum principle that combined mechanism (3), with the lowest load factor, is the correct collapse mechanism and that the load factor against collapse is 1.92.

However, in more complex cases, with more than one bay or more than one storey it is not possible to be so confident that the correct mechanism has been obtained. It is therefore regarded as essential to draw an admissible bending moment diagram at collapse in order to confirm the result using the uniqueness theorem. The known situation at collapse is as shown in Fig. 2.19.

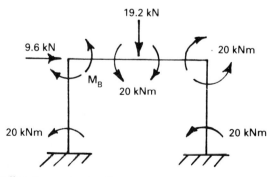

Fig. 2.19 Bending moments at collapse.

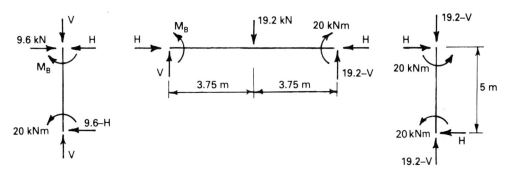

Fig. 2.20 Equilibrium of individual members at collapse.

In this instance, but not generally, the frame is statically determinate at collapse and the single unknown bending moment M_B can be determined in one of two ways.

The first method is to consider the equilibrium of the individual members as shown in Fig. 2.20. The requirements of horizontal and vertical equilibrium have been satisfied by inserting the forces in the diagram and the three unknowns, H, V and M_B can be readily determined from moment equilibrium as follows:

From moment equilibrium of the right-hand stanchion

$$H = 8\,\text{kN}$$

From moment equilibrium of the left-hand stanchion

$$M_B + 5(9.6 - H) - 20 = 0$$

$$\therefore \quad M_B = 20 - 5 \times 1.6 = 12\,\text{kNm}$$

For completeness, from moment equilibrium of the beam

$$7.5V + 20 - M_B - 19.2 \times 3.75 = 0$$

$$\therefore \quad V = \frac{72 + 12 - 20}{7.5} = 8.53\,\text{kN}$$

The bending moment diagram at collapse is therefore as shown in Fig. 2.21 (kNm) and the correct collapse mechanism is confirmed.

In the second method, the value of M_B can be determined much more quickly by using virtual work. However, when using this method it is necessary to consider carefully the signs of the internal work terms on the right-hand side of the equation as it can no longer be assumed that they are automatically positive. We assume a virtual mechanism movement of the beam inserting the known conditions at collapse. This is standard use of the virtual work method whereby an arbitrary

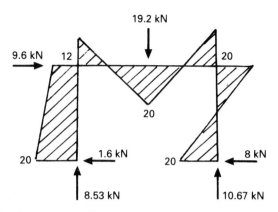

Fig. 2.21 Bending moment diagram at collapse.

but compatible system of displacements is combined with the real internal and external force system in order to obtain an equilibrium condition. Here the displacement system is chosen to be a rigid-link mechanism as shown in Fig. 2.22.

The virtual work equation is

$$19.2 \times 3.75\theta = M_B\theta + 20(2\theta + \theta)$$

i.e. $M_B = 19.2 \times 3.75 - 60 = 12.0\,\text{kNm}$ as before.

Note.   If there was good reason to believe the combined mechanism (3) to be the correct collapse mechanism, it would have been possible to proceed directly to this mechanism and to confirm the result by drawing the bending moment diagram without considering all other mechanisms.

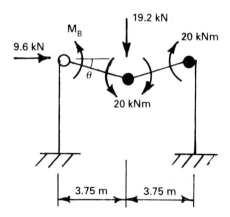

Fig. 2.22 Determination of M_B using virtual work.

2.4   Partial and overcomplete collapse

The frame considered in Example 2.3 above collapsed in a combined mechanism with four plastic hinges and was statically determinate at collapse. The frame was a simple rectangular frame with three degrees of redundancy. This simple observation can be generalised into the following definitions:

A complete collapse mechanism is such that the structure is statically determinate at collapse and contains one more plastic hinge than the degree of redundancy of the structure.

A mechanism with more plastic hinges than are necessary for complete collapse is termed *overcomplete*.

A mechanism with less plastic hinges than are necessary for complete collapse is termed a *partial collapse mechanism*.

These definitions are of considerable practical significance and the following points should be carefully noted:

❑ An overcomplete mechanism has more than one degree of freedom and should *never* be analysed. Overcomplete mechanisms are invariably the result of two or more alternative mechanisms arising simultaneously and it is the component mechanisms that must be considered.

❑ Complete collapse is *not* essential. The beam mechanism in Example 2.3 was an example of partial collapse and many other structures collapse without forming sufficient plastic hinges for complete collapse. Figure 2.23 shows a more comprehensive example of partial collapse in which:

Degree of redundancy $\quad r = 9$
Number of plastic hinges $\quad n = 6$

$n < r + 1$
$\therefore$ partial collapse

❑ In a statically determinate structure, the first plastic hinge causes complete collapse.

❑ When a frame suffers partial collapse it is not possible to draw a unique bending moment diagram at collapse. It is, however, still possible to satisfy the uniqueness theorem. This is illustrated by Example 2.4 below.

Example 2.4   The frame shown in Fig. 2.24 has a uniform full plastic moment of 40 kNm. The loads shown are unfactored, find the load factor λ against collapse.

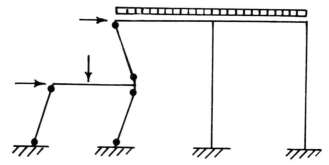

Fig. 2.23 Example of partial collapse.

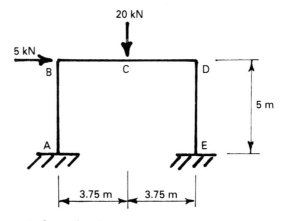

Fig. 2.24 Frame to be analysed.

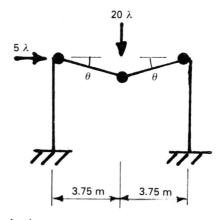

Fig. 2.25 Beam mechanism.

From comparison with Example 2.3, it is reasonably obvious that the correct collapse mechanism is a beam mechanism and that the load factor can be readily calculated using Fig. 2.25.

$$20\lambda \times 3.75\theta = 40(\theta + 2\theta + \theta)$$

$$\lambda = \frac{160}{75} = 2.13$$

The bending moments in the beam are fixed by the collapse mechanism but, in order to draw the bending moment diagram at collapse, it is necessary to know the bending moments at both A and E. However, the frame is not statically determinate at collapse, and these moments cannot be determined uniquely. It is, nevertheless, possible to determine an equilibrium relationship that they must satisfy and this is best done by virtual work.

Consider the virtual sway mechanism shown in Fig. 2.26 under the collapse conditions determined above:

$$10.67 \times 5\theta = M_A\theta - 40\theta - 40\theta + M_E\theta$$

$$\therefore \quad M_A + M_E = 53.3 \, \text{kNm}$$

There are numerous bending moment distributions which satisfy this condition and also satisfy the yield condition. The bending moment diagram at collapse can therefore lie anywhere between the two extremes shown in Fig. 2.27. Despite the fact that a unique bending moment diagram at collapse cannot be drawn, the presence of any bending moment diagram satisfying the conditions of equilibrium, mechanism and yield is sufficient to satisfy the uniqueness theorem and to confirm the above result.

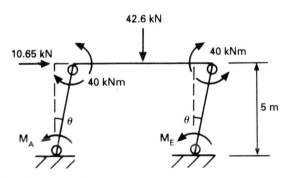

Fig. 2.26 Virtual sway mechanism.

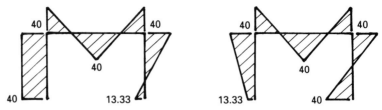

Fig. 2.27 Limiting bending moment diagrams at collapse (kNm).

2.5   Further considerations in the analysis of portal frames

Practical frames are generally subject to uniformly distributed loads, rather than the point loads that were used to illustrate some basic principles in Sections 2.3 and 2.4. Furthermore, frames may have more than one span or more than one storey. In such cases, it is generally too complicated to work with exact positions for the plastic hinges which form in beams below distributed loads and it is sufficient to work with approximate hinge positions provided that the implications are fully understood.

The essential consideration is that, unless the solution obtained is corrected using the static theorem, an unsafe answer will be obtained. The principles involved are illustrated by the following example.

Example 2.5   Figure 2.28 shows the loads and dimensions of a pinned based frame. The stanchions have a full plastic moment of 20 kNm and the beams have a full plastic moment of 36 kNm. Obtain an estimate of the load factor against collapse.

In the first instance, it is assumed that plastic hinges within the span of the beams form at mid-span. All reasonable mechanisms are considered, as shown in Figs 2.29(a)–(e), and the one with the lowest

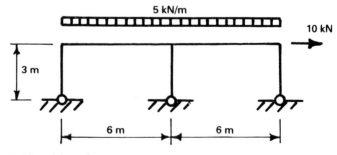

Fig. 2.28 Pinned base frame.

load factor is chosen in accordance with the kinematic theorem. The bending moment diagram for this frame is then drawn and the implications of the assumed plastic hinge positions examined.

It should be carefully noted that where two unequal members meet, as in the eaves connections between the stanchions and the beam, the bending moment in both members is the same and the plastic hinge will always form in the weaker member.

(a)   *Beam mechanism*

$$30\lambda \times 1.5\theta = 20\theta + 36 \times 3\theta$$

$$\therefore \quad \lambda = 2.844$$

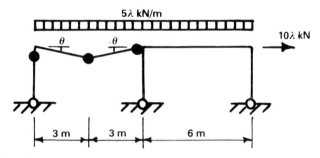

Fig. 2.29(a) Beam mechanism.

(b)   *Sway mechanism*

$$10\lambda \times 3\theta = 20 \times 3\theta$$

$$\therefore \quad \lambda = 2.0$$

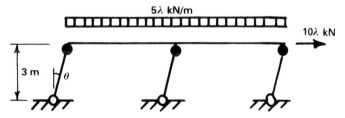

Fig. 2.29(b) Sway mechanism.

(c)   *Combined mechanism No. 1*

$$30\lambda \times 1.5\theta + 10\lambda \times 3\theta = 20 \times 2\theta + 36 \times 3\theta$$

$$(45 + 30)\lambda = 148$$

$$\therefore \quad \lambda = 1.973$$

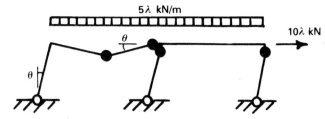

Fig. 2.29(c) Combined mechanism No. 1.

(d) *Combined mechanism No. 2*

$$30\lambda \times 1.5\theta + 10\lambda \times 3\theta = 20 \times 3\theta + 36 \times 3\theta$$

$$(45 + 30)\lambda = 168$$

$$\therefore \quad \lambda = 2.240$$

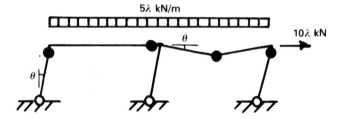

Fig. 2.29(d) Combined mechanism No. 2.

(e) *Combined mechanism No. 3*

$$2 \times 30\lambda \times 1.5\theta + 10\lambda \times 3\theta = 20 \times 2\theta + 36 \times 6\theta$$

$$(90 + 30)\lambda = 256$$

$$\therefore \quad \lambda = 2.133$$

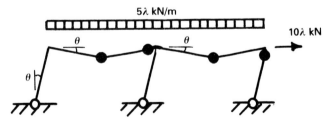

Fig. 2.29(e) Combined mechanism No. 3.

Thus, combined mechanism No. 1 is critical with a load factor against collapse of 1.973. There are four plastic hinges in the collapse mechanism, which is one more than the degree of redundancy, so that the frame is statically determinate at collapse.

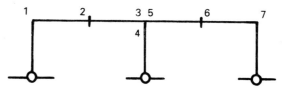

Fig. 2.30 Critical sections.

It is now necessary to draw the bending moment diagram at collapse and, in order to do this, it is necessary to determine the bending moment at the seven critical sections numbered in Fig. 2.30.

Before we proceed any further, it is necessary to pause for a moment to consider the number of independent equilibrium equations which are available to draw this bending moment diagram. There is a simple rule for this which can be illustrated with reference to the portal frame considered in the previous section. In order to draw the bending moment diagram, it is necessary to determine the bending moment at the five numbered locations in Fig. 2.31. The frame is three degrees redundant and can be made statically determinate by introducing a cut at one base. This requires the insertion of the three redundant forces, M, H and V.

The bending moments at the five critical sections can now be written down in terms of the known applied loads, W_1 and W_2, together with the three unknowns M, H and V, e.g.

$$M_1 = W_1h + W_2L - M - 2LV$$

$$M_2 = W_2L - M + hH - 2LV$$

etc.

Eliminating M, H and V from the above equations leaves two independent equilibrium equations in the critical section bending

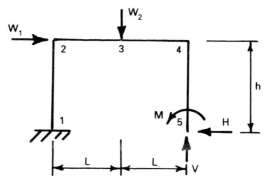

Fig. 2.31 Redundant forces in a single-bay frame.

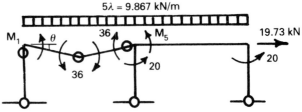

Fig. 2.32 Left-hand beam virtual mechanism.

moments. Regardless of the exact form of these equations, the number of them is precisely determined by the above argument. It is easy to see that this argument can be generalised.

Let $n =$ number of critical sections (where the bending moment must be determined in order to draw the bending moment diagram)

$r =$ degree of redundancy

$n - r =$ number of independent equilibrium equations.

Returning to the two-bay example under consideration, the frame has seven critical sections and is three degrees redundant so that there are $7 - 3 = 4$ independent equilibrium relationships which must be satisfied by the bending moment diagram. One of these has been used above in obtaining the solution for the mechanism (c) under examination. This leaves three equations available for continuing the analysis which can now proceed as follows.

(1) Determine M_1 using a virtual mechanism in the left-hand beam as shown in Fig. 2.32.

$9.867 \times 6 \times 1.5\theta = M_1\theta + 3 \times 36\theta$

$M_1 = -19.2\,\text{kNm}$

(2) Determine M_5 from equilibrium of the central joint.

$M_5 = 36 - 20 = 16\,\text{kNm}$

(3) Determine M_6 using a virtual mechanism in the right-hand beam as shown in Fig. 2.33.

$9.867 \times 6 \times 1.5\theta = (16 + 20)\theta + 2M_6\theta$

$M_6 = 26.4\,\text{kNm}$

The bending moment diagram at collapse can now be drawn as shown in Fig. 2.34.

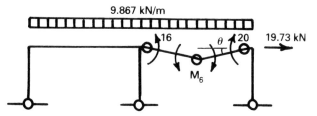

Fig. 2.33 Right-hand beam virtual mechanism.

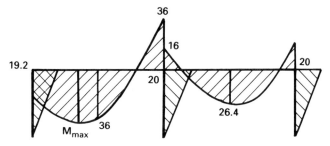

Fig. 2.34 Bending moment diagram at collapse.

It is obvious that the full plastic moment of 36 kNm is exceeded within the left-hand beam. It is therefore necessary to determine the position and magnitude of the maximum bending moment within this span. This is elementary structural mechanics. The relevant forces are shown in Fig. 2.35 and it follows from simple equilibrium that:

$$R_A = \frac{1}{6}(9.867 \times 6 \times 3 - 36 - 19.2) = 20.4\,\text{kN}$$

$$M_x = 19.2 + 20.4x - 9.867\,\frac{x^2}{2}\ \text{kNm}$$

The maximum bending moment is where the shear force is zero, i.e. at

$$x = \frac{20.4}{9.867} = 2.0675\,\text{m}$$

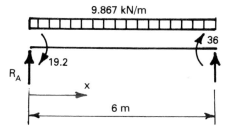

Fig. 2.35 Equilibrium of left-hand beam.

so that

$$M_{max} = 19.2 + 20.4 \times 2.0675 - 9.867 \times \frac{2.0675^2}{2} = 40.29 \text{ kNm}$$

This is significantly in excess of the full plastic moment of 36 kNm. However, using the static theorem, it can be stated that a lower bound on the load factor is given by

$$\lambda = 1.973 \times \frac{36.0}{40.29} = 1.763$$

so that

$$1.763 \le \lambda \le 1.973$$

If this is not sufficiently accurate, the analysis can be repeated with the more precise plastic hinge position obtained above.

2.5.1   The implications of partial collapse in more complex frames

At first sight it might appear that the procedure illustrated in Example 2.5 could break down if the critical collapse mechanism turns out to be a case of partial collapse so that the frame is not statically determinate at collapse. This is not the case because, in order to satisfy the uniqueness theorem, it is not necessary to draw the precise bending moment diagram, merely one that satisfies the yield criterion and all the independent equilibrium equations. As the requisite number of equilibrium equations can be readily determined as shown in the previous section $(n - r)$ the problem of partial collapse is not significantly more difficult than the case of complete collapse considered in Example 2.5. Example 2.6 illustrates this point.

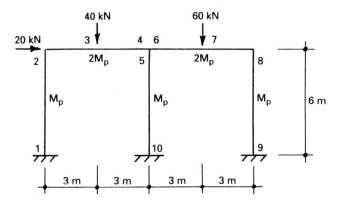

Fig. 2.36 Frame for Example 2.6.

Example 2.6   Partial collapse.

It is required to design the frame shown in Fig. 2.36 for the factored loads shown with the full plastic moment of the beam twice that of the stanchions.

Because, in a severe case of partial collapse, it is useful not merely to identify the critical mechanism (in this case a beam mechanism) but also other near-critical mechanisms, we will again systematically consider all possible mechanisms. As by now the work equation method should be familiar, these will be analysed with the minimum of explanation. Note, however, that the stanchions are weaker than the beams and that the eaves hinges will always form in the weaker stanchions.

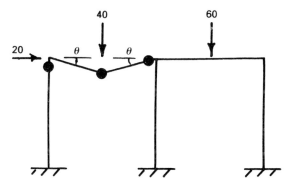

Fig. 2.37 Left-hand beam mechanism.

Left-hand beam mechanism (Fig. 2.37)

$$120\theta = 7M_p\theta$$

$$\therefore \quad M_p = 17.2\,\text{kNm}$$

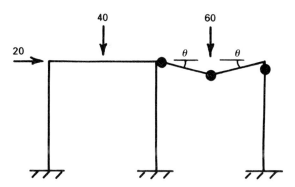

Fig. 2.38 Right-hand beam mechanism.

Right-hand beam mechanism (Fig. 2.38)

$$180\theta = 7M_p\theta$$

$$\therefore \quad M_p = 25.7\,\text{kNm}$$

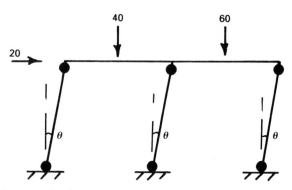

Fig. 2.39 Sway mechanism.

Sway mechanism (Fig. 2.39)

$$120\theta = 6M_p\theta$$

$$\therefore \quad M_p = 20.0\,\text{kNm}$$

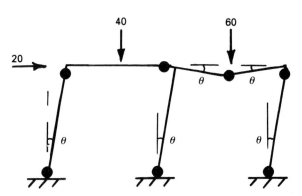

Fig. 2.40 Combined mechanism No. 1.

Combined mechanism No. 1 (Fig. 2.40)

$$300\theta = 12M_p\theta$$

$$\therefore \quad M_p = 25.0\,\text{kNm}$$

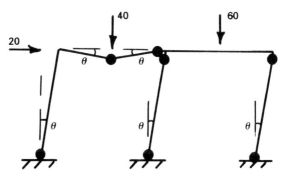

Fig. 2.41 Combined mechanism No. 2.

Combined mechanism No. 2 (Fig. 2.41)

$$240\theta = 11M_p\theta$$

$$\therefore \quad M_p = 21.8\,\text{kNm}$$

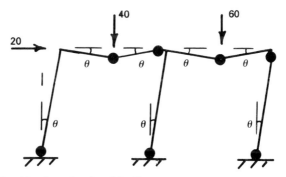

Fig. 2.42 Combined mechanism No. 3.

Combined mechanism No. 3 (Fig. 2.42)

$$420\theta = 17M_p\theta$$

$$\therefore \quad M_p = 24.7\,\text{kNm}$$

It is therefore concluded that the correct collapse mechanism is a partial collapse involving only the right-hand beam with

$$M_p = 25.7\,\text{kNm}$$

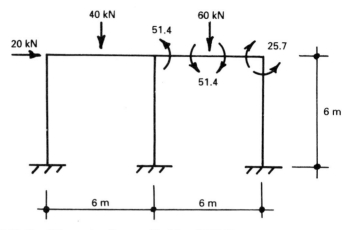

Fig. 2.43 Conditions at collapse with $M_p = 25.7$ kNm.

This can be confirmed by finding *any* bending moment which satisfies:

- ❑ all independent equilibrium equations;
- ❑ the mechanism condition found above;
- ❑ the yield condition (the full plastic moment is nowhere exceeded).

Because the mechanism found has only three plastic hinges compared with the seven required for complete collapse, this is tricky. The information available at this stage is summarised in Fig. 2.43.

We first note that there are $r = 10$ critical sections where plastic hinges may form (numbered on Fig. 2.36) and that the degree of redundancy $n = 6$. It follows that there are $n - r = 4$ independent equilibrium equations which must be satisfied. We have used one of these in determining $M_p = 25.7$ kNm so that there are only three more to be satisfied in drawing the bending moment diagram.

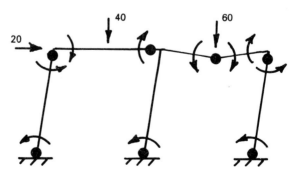

Fig. 2.44 Plastic hinges in the next most critical mechanism.

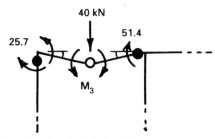

Fig. 2.45 Virtual mechanism in left-hand beam.

We now note that the next most critical collapse mechanism is combined mechanism No. 1 with the plastic hinge positions shown in Fig. 2.44. As this mechanism required $M_p = 25.0\,\text{kNm}$ compared with the final solution of $M_p = 25.7\,\text{kNm}$, it may be expected that most, though not all, of the plastic hinges shown in Fig. 2.44 will be present at collapse. This observation removes most of the guesswork in establishing a statically admissible bending moment distribution which satisfies the yield condition.

If we first assume plastic hinges at both ends of the left-hand beam, as suggested by Fig. 2.44, we can determine the bending moment M_3 at the centre of this beam from the equilibrium of the beam. The easiest way to do this is by virtual work. From Fig. 2.45, observing the signs of the bending moments:

$$40 \times 3\theta = (-25.7 + 51.4 + 2M_3)\theta$$

$$\therefore \quad M_3 = 47.15\,\text{kNm}$$

We now have equal and opposite bending moments in the beam on either side of the central joint. Equilibrium at this joint indicates that the bending moment at critical section number 5 must be zero as indicated in Fig. 2.46.

We have now used three of the four available equilibrium equations. The final equilibrium equation can be obtained by considering a sway

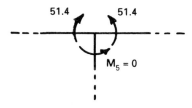

Fig. 2.46 Joint equilibrium.

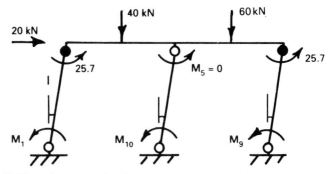

Fig. 2.47 Virtual sway mechanism.

virtual work mechanism as shown in Fig. 2.47. The equilibrium equation obtained in this way includes the remaining unknown bending moments at the column bases, thus:

$$20 \times 6\theta = (25.7 + 0 + 25.7 + M_1 + M_{10} + M_9)\theta$$

$$\therefore \quad M_1 + M_{10} + M_9 = 68.6\,\text{kNm}$$

Although this equation does not allow these bending moments to be determined uniquely, it does allow a reasonable guess to be made, e.g.

$$M_9 = M_{10} = 25.7\,\text{kNm}, \quad M_1 = 17.1\,\text{kNm}$$

and, on this basis, a bending moment diagram can be drawn as shown in Fig. 2.48.

Although this diagram is not unique, it satisfies all the available equilibrium conditions and does not violate the yield condition. It therefore provides valid confirmation that the correct collapse mechanism has been obtained. It is also sufficiently accurate to enable the

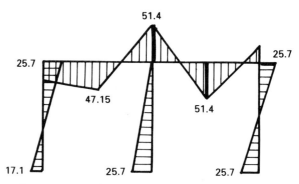

Fig. 2.48 Bending moment distribution at collapse.

member stability checks to be carried out as will be described in the next chapter. Indeed, Fig. 2.48 is very close to the exact solution determined by elastic-plastic computer analysis (which is the only way to obtain a more accurate bending moment distribution).

2.6   Frames with sloping members

When a frame includes one or more sloping members, no new considerations are introduced other than the possibility of alternative collapse mechanisms and a certain additional difficulty in determining the rotations and displacements in the virtual work equation. These additional factors will be illustrated by means of a further example.

Example 2.7   The frame shown in Fig. 2.49 has a uniform section with a full plastic moment of 300 kNm. Find the load factor at collapse.

As no loads are applied within the lengths of the members, plastic hinges can only form at sections ABCDE because the bending moment diagram must be linear between these points. There are, therefore, only three valid mechanisms with hinges at these sections and these will be considered in turn. By the minimum principle, the correct collapse mechanism will be the one having the lowest load factor.

Mechanism No. 1 (Fig. 2.50)
Figure 2.50 shows a valid collapse mechanism with plastic hinges at B, C, D and E.

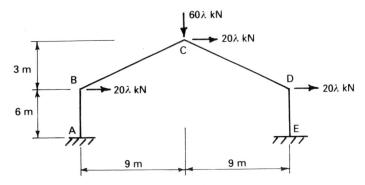

Fig. 2.49 Pitched roof frame to be analysed.

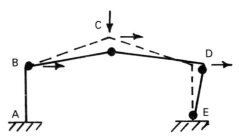

Fig. 2.50 Collapse mechanism No. 1.

Using the virtual work method, it is necessary to relate all of the rotations and displacements to a single rotational variable θ. There is a convenient method for doing this that is applicable to all problems of this type and this involves a procedure that will be more familiar to mechanical engineers. The collapse mechanism involves three moving bars and the first step is to determine the instantaneous centre of rotation of the middle bar CD. This point is I_{CD} in Fig. 2.51.

I_{CD} is easily located by noting that, for a small initial movement of the mechanism, C moves to C′ along a line at right angles to BC. The instantaneous centre of rotation of CD must therefore be somewhere along BC produced. Similarly, D moves to D′ along a line at right angles to ED so that I_{CD} must be on ED produced. I_{CD} must therefore lie at the meeting point of BC and ED produced and member CD can move from its original position to its new position C′D′ by a pure rotation about I_{CD}.

We now consider a movement of the mechanism characterised by a *small* rotation θ of member CD about I_{CD}. The other rotations shown

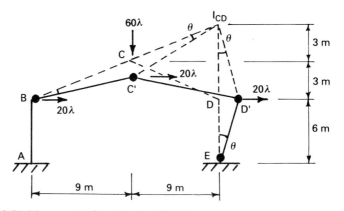

Fig. 2.51 Movement of mechanism No. 1.

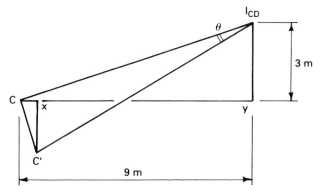

Fig. 2.52 Horizontal and vertical components of movement at C.

on Fig. 2.51 follow directly by similar triangles. The virtual work equation is therefore

$$20\lambda \times 3\theta + 60\lambda \times 9\theta + 20\lambda \times 6\theta = 300(\theta + 2\theta + 2\theta + \theta)$$

$$\text{i.e.}\quad \lambda = 2.50$$

Readers who find difficulty in visualising the displacements at C may find it easier to draw an additional diagram, as shown in Fig. 2.52.

Triangle CC'X is similar to triangle I_{CD}CY though rotated by 90° and scaled by θ. It follows directly that the horizontal component of displacement at C is 3θ and the vertical component is 9θ as included in the above virtual work equation.

Mechanism No. 2 (Fig. 2.53)
Here, Fig. 2.53 shows an alternative valid collapse mechanism with plastic hinges at A, C, D and E.

The instantaneous centre of rotation I_{CD} of the middle of the three moving bars is found by noting that ABC is a rigid arm rotating about A so that C moves to C' at right angles to the broken line joining A to C. I_{CD} therefore lies on AC produced. It also lies, as before, on ED produced so that the position of I_{CD} lies at the intersection of AC and ED as shown in Fig. 2.53. The mechanism movement is again defined by a small rotation θ of member CD about I_{CD} and the remaining rotations of ABC and ED follow as shown.

The virtual work equation for this mechanism is therefore

$$20\lambda \times 6\theta + 20\lambda \times 9\theta + 60\lambda \times 9\theta + 20\lambda \times 12\theta$$

$$= 300(\theta + 2\theta + 3\theta + 2\theta)$$

$$\text{i.e.}\quad \lambda = 2.22$$

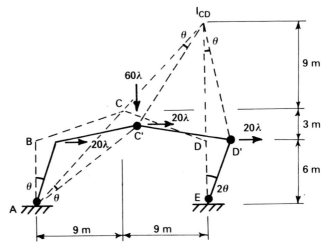

Fig. 2.53 Movement of mechanism No. 2.

Mechanism No. 3 (Fig. 2.54)
The third possible collapse mechanism is the sway mechanism shown in Fig. 2.54 with plastic hinges at A, B, D and E. The rafter BCD sways bodily sideways without any vertical movement so that the work equation is comparatively simple and can be written down directly.

$$\text{Thus} \quad 3 \times 20\lambda \times 6\theta = 300 \times 4\theta$$

$$\text{i.e.} \quad \lambda = 3.33$$

It therefore follows that the correct collapse mechanism is No. 2 which has the lowest load factor namely λ equal to 2.22.
In order to complete the problem, it is necessary to determine the bending moment at B at collapse and to draw the bending moment diagram. This too is best done by using the method of virtual work.

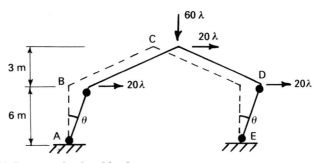

Fig. 2.54 Sway mechanism No. 3.

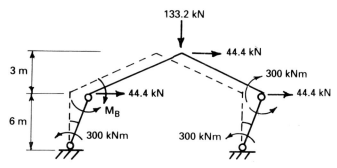

Fig. 2.55 Virtual sway mechanism.

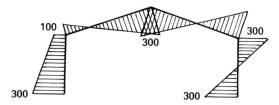

Fig. 2.56 Bending moment diagram at collapse.

Consider a virtual sway mechanism and insert the load factor and bending moments at collapse found for mechanism No. 2, as shown in Fig. 2.55.

$$3 \times 44.4 \times 6\theta = 300 \times 3\theta + M_B\theta$$

$$M_B = -100\,\text{kNm}$$

The minus sign indicates that M_B in Fig. 2.55 is drawn in the wrong direction so that the bending moment diagram at collapse is as shown in Fig. 2.56 (kNm).

2.6.1 *Pitched roof portal frames using free and reactant bending moment diagrams*

Pitched roof portal frames are frequently analysed using a semi-graphical method that involves combining free and reactant bending moment diagrams in much the same way that was demonstrated earlier for beams. This method avoids the complexities of instantaneous centres of rotation and has other benefits which will be apparent when haunched frames and more complex loading conditions are dealt with later. However, for many frames, the choice between the two alternative methods is mainly a matter of personal preference.

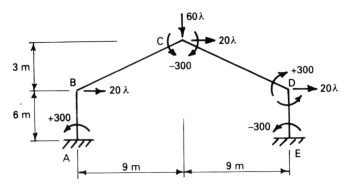

Fig. 2.57 Alternative analysis of a pitched roof portal frame.

The method will first be illustrated by repeating the analysis of the critical mechanism in Example 2.7.

Example 2.8   Alternative analysis of mechanism No. 2 of Example 2.7.

This mechanism is shown in Fig. 2.57 which includes all the information necessary for this method of analysis.

In this method of analysis, it is convenient to use a sign convention in which bending moments causing tension on the outside of the frame are considered positive and the moments shown in Fig. 2.57 have been given signs on this basis.

This frame can be made statically determinate by cutting it at the apex. The actual conditions at any state of loading can then be considered to be the sum of the two cases shown in Fig. 2.58(a) and (b).

Figure 2.58(a) shows the actual loads taken on the statically determinate (cut) frame and gives rise to the free bending moment diagram shown in Fig. 2.59(a). The cut at the apex releases three forces M, H and V which give rise to the reactant bending moment diagram

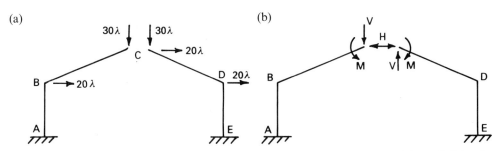

Fig. 2.58 Internal forces in frame as the sum of free and reactant systems: (a) applied loads on statically determinate frame; (b) forces released by cutting frame.

shown in Fig. 2.59(b). Provided that values of the four unknowns, M, H, V and λ have been determined, the actual bending moment diagram for the structure under any condition of loading can always be obtained by adding together these two component diagrams. In order to simplify the presentation, the bending moments are shown on a horizontal base by 'unwrapping' the frame. Positive bending moments put the inner flange into compression so that the free bending moment diagram is mainly on the positive side of the horizontal axis and the reactant moment diagram on the negative side.

The determination of M, H, V and λ requires four equations. Here, these four equations are obtained by equating the bending moment at the four plastic hinge positions shown in Fig. 2.53 to the full plastic moment of resistance of the members (300 kNm) while observing the consistent sign convention described above. Thus:

$$\text{at A:} \quad 390\lambda - M - 9H + 9V = 300$$

$$\text{at C:} \quad 0 \ - M = -300$$

$$\text{at D:} \quad 210\lambda - M - 3H - 9V = 300$$

$$\text{at E:} \quad -30\lambda - M - 9H - 9V = -300$$

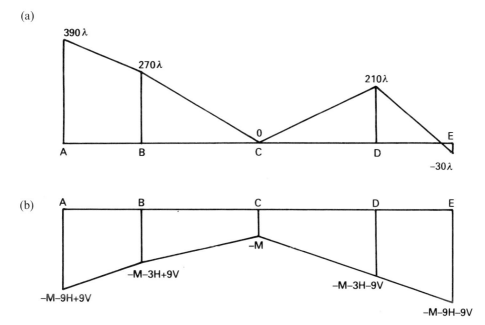

Fig. 2.59 Free and reactant bending moment diagrams: (a) free bending moment diagram (kNm); (b) reactant bending moment diagram.

These four simultaneous equations can be readily solved to give

$\lambda = 2.222$
$M = 300\,\text{kNm}$
$H = 11.11\,\text{kN}$
$V = -18.52\,\text{kN}$

The bending moment M_B at B also follows immediately by substituting the above values into the following equation:

$$M_B = 270\lambda - M - 3H + 9V$$

i.e. $M_B = 100.0\,\text{kNm}$

The bending moment diagram can then be drawn to verify that the correct collapse mechanism has been analysed. This has been given previously as Fig. 2.56.

2.6.2   Pitched roof portal frames subject to uniformly distributed load

The most important single case with which designers of low-rise steel structures may be confronted is probably the pitched roof portal frame subject to a vertical uniformly distributed load. Frames of this type are almost always designed by plastic theory and wind loads rarely govern the design so that this is a case of frequent practical importance. For this reason it will be considered in some detail, first using the virtual work method and then using the alternative semi-graphical method.

Figure 2.60 shows a typical haunched portal frame and the most usual symmetrical collapse mechanism. There are a number of general points to note.

❑ Strictly speaking, this mechanism is overcomplete and should not be analysed. However, if the constraint of symmetry is applied so that the apex moves down vertically without rotating, this mechanism becomes 'complete'.
❑ The position of the rafter hinges is unknown. In the example which follows, these plastic hinge positions will be treated as variables and the critical positions found by calculus. In practice, the uniformly distributed load from the roof is usually applied to the frame as a series of point loads through purlins. When this is the case, the rafter hinges invariably form under the first or second purlin down from

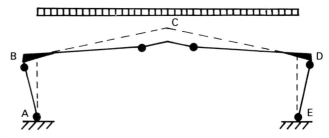

Fig. 2.60 Symmetrical collapse mechanism.

the apex. An alternative procedure is therefore to consider the hinges in these positions from the outset. The analysis is then similar to, and simpler than, the more general approach which follows.

❑ The method is essentially the same whether the frame has pinned or fixed feet.

The design of a typical haunched frame will now be illustrated by means of an example.

Example 2.9   The loads shown in Fig. 2.61 are factored loads. The frame is to be designed on the assumption that the full plastic moment of the stanchion is 1.5 times the full plastic moment of the rafters and that the haunch is of sufficient length to ensure that the plastic hinge at the eaves forms in the stanchion below the haunch and not in the rafter.

The plastic collapse mechanism is shown in Fig. 2.62. For a solution using the virtual work method the following points should be noted.

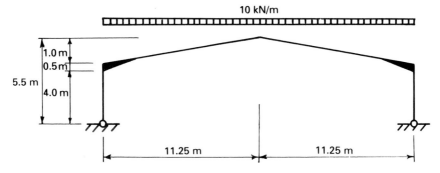

Fig. 2.61 Symmetrical frame to be designed.

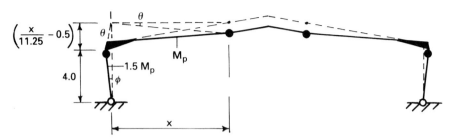

Fig. 2.62 Collapse mechanism for a symmetrical pitched roof portal frame.

❑ The position of the rafter hinge is unknown. The horizontal distance x from the stanchion to this hinge will therefore be treated as a variable and its value determined as part of the solution.

❑ Symmetry conditions require that the apex of the frame moves down vertically without lateral movement or rotation. It follows that the whole centre section of the rafter descends bodily.

❑ The central bar of the mechanism between the plastic hinges is a rigid section which rotates about the instantaneous centre denoted by I in Fig. 2.62. The rotation of this section is denoted θ.

The rotation ϕ of the stanchion follows by considering the horizontal displacement of the plastic hinge in the stanchion which can be expressed in terms of θ or ϕ:

$$4\phi = \left(\frac{x}{11.25} + 0.5\right)\theta$$

i.e. $\quad \phi = \left(\dfrac{x}{45} + \dfrac{1}{8}\right)\theta$

The virtual work equation can then be written down for half of the frame:

$$10x\frac{x\theta}{2} + 10(11.25 - x)x\theta = 1.5M_p(\theta + \phi) + M_p\theta$$

i.e. $\quad 5x^2 + 112.5x - 10x^2 = 2.5M_p + 1.5M_p\left(\dfrac{x}{45} + \dfrac{1}{8}\right)$

i.e. $\quad M_p = \dfrac{112.5x - 5x^2}{2.6875 + \dfrac{x}{30}}$

The critical value of M_p is the maximum value as the rafter hinge moves.

This can be found by calculus, the requirement being

$$\frac{dM_p}{dx} = 0$$

i.e. $\left(2.6875 + \dfrac{x}{30}\right)(112.5 - 10x) - (112.5x - 5x^2)\dfrac{1}{30} = 0$

i.e. $x^2 + 161.25x - 1814.06 = 0$

i.e. $x = \dfrac{-161.25 \pm \sqrt{161.25^2 + 4 \times 1814.06}}{2}$

i.e. $x = 10.56\,\text{m}$ or $-171.8\,\text{m}$

Of these two solutions, the only one within the rafter is $x = 10.56$ m. The required value of M_p is therefore

$$M_p = \frac{112.5 \times 10.56 - 5 \times 10.56^2}{2.6875 + \dfrac{10.56}{30}}$$

$$= 207.4\,\text{kNm}$$

It follows that a suitable design for the frame could be obtained by choosing members with full plastic moments close to the following values:

stanchion   311.1 kNm
rafter   207.4 kNm

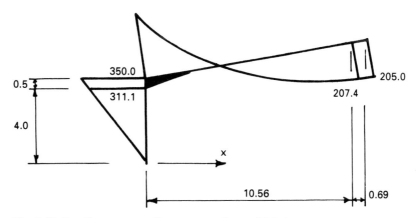

Fig. 2.63 Bending moment diagram at collapse (kNm).

As the frame is statically determinate at collapse, the bending moment diagram can be readily drawn and the required haunch length determined as shown in Fig. 2.63.

The eaves moment is given by $311.1 \times \dfrac{4.5}{4.0} = 350.0\,\text{kNm}$

The equation for the bending moment in the rafter can be obtained by considering the length between the eaves and the plastic hinge position where the moment is 207.4. Thus:

$$M_x = -350.0 + \left(\frac{350.0 + 207.4}{10.56}\right) x + 10x \left(\frac{10.56 - x}{2}\right)$$

$$= -350.0 + 105.58x - 5x^2$$

when $x = 11.25\,\text{m}$, the bending moment at the apex is given by

$$M_{11.25} = -350.0 + 1187.8 - 632.8 = 205.0\,\text{kNm}$$

The minimum length of the haunch is given by the condition that the end of the haunch remains elastic. If a shape factor of 1.15 is assumed, this gives

$$M_x = -\,207.4/1.15 = -180.3\,\text{kNm}$$

i.e. $-5x^2 + 105.58x - 350.0 = -180.3$

i.e. $x^2 - 21.12x + 33.94 = 0$

i.e. $x = +\dfrac{21.12 \pm \sqrt{21.12^2 - 4 \times 33.94}}{2}$

i.e. $x = 1.75\,\text{m}$ or $19.37\,\text{m}$

The minimum haunch length is therefore 1.75 m.

Note: The method described above is equally valid if the frame has fixed feet. There is then merely an extra term of $1.5M_p\phi$ on the right-hand side of the virtual work equation.

The work equation method becomes rather more complicated if either the frame or loading are not symmetrical. For this reason, many engineers prefer to use semi-graphical methods for the plastic design of pitched roof portal frames. In order to further illustrate the principles involved, the above example will be reworked using the semi-graphical method.

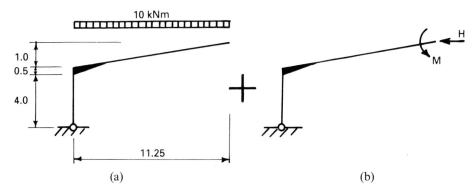

Fig. 2.64 Internal forces as the sum of free and reactant systems: (a) applied loads on the statically determinate structure; (b) forces released by cutting the frame.

Example 2.10 Alternative analysis of Example 2.9 using free and reactant bending moment diagrams.

The frame is again made statically determinate by introducing a cut at the apex so that the actual internal forces under any conditions of loading are given by the sum of the two situations shown in Figs 2.64(a) and (b). Here it should be noted that because both the frame and loading are symmetrical it is only necessary to consider one half of the structure and there is no need to include a vertical force at the cut in Fig. 2.64(b).

The bending moment diagrams arising from the two force systems shown in Fig. 2.64 can be readily drawn as shown in Fig. 2.65. Once again, the diagrams are drawn on a horizontal base by 'unwrapping' the frame.

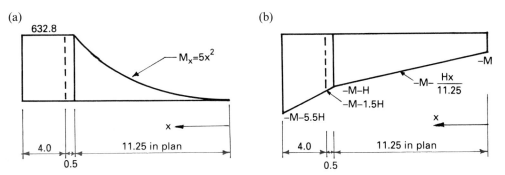

Fig. 2.65 Free and reactant bending moment diagrams: (a) free bending moment diagram (kNm); (b) reactant moment diagram.

As there is a pinned base, the bending moment there must be zero so that, whatever conditions exist in the frame, the following equation must always be satisfied:

$$632.8 - M - 5.5H = 0$$

The remaining equations necessary to complete the solution follow from the plastic collapse mechanism. The only feasible mechanism is shown in Fig. 2.62 and if it is again assumed that the full plastic moment of the rafter is M_p and that of the stanchion is $1.5M_p$ we have

$$632.8 - M - 1.5H = 1.5M_p$$

$$5x^2 - M - \frac{Hx}{11.25} = -M_p$$

Eliminating the redundant forces M and H gives

$$M_p = \frac{632.8 - 5x^2}{3.0625 - \dfrac{x}{30}}$$

Following the same procedure that was described in Example 2.9, the critical hinge position is given by

$$\frac{\mathrm{d}M_p}{\mathrm{d}x} = 0$$

i.e. $x = 0.690\,\mathrm{m}$

and the requisite value of M_p is again

$$M_p = 207.4\,\mathrm{kNm}$$

which is, of course, precisely the same solution that was obtained previously.

However, the semi-graphical method described above is rather more versatile than the virtual work method when adjustments to accommodate such practicalities as the choice of discrete member sizes from section tables and variations in the dimensions of the haunch have to be made.

If we make the reasonable assumption that the plastic hinge in the rafter forms at the first purlin point distant (say) 1.5 m from the apex, the basic equations can be simplified. We also take advantage of the fact that the bending moment at the hinged base must always be zero to eliminate M from the equations but retain complete freedom of choice of the values for the full plastic moments of the rafter and stanchion as

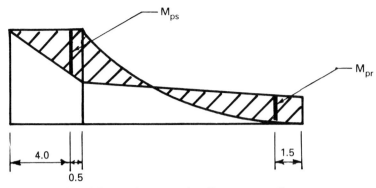

Fig. 2.66 Combined free and reactant bending moment diagram.

M_{pr} and M_{ps} respectively. Then, the three equations

$$632.8 - M - 5.5H = 0$$

$$632.8 - M - 1.5H = M_{ps}$$

$$11.25 - M - \frac{H}{7.5} = -M_{pr}$$

reduce to

$$4H = M_{ps}$$

$$621.55 - 5.367H = M_{pr}$$

The bending momemt diagram is as shown in Fig. 2.66 and adjusting it according to the above equations in order to accommodate alternative values of M_{ps} and M_{pr} is extremely simple.

2.6.3 Pitched roof portal frames subject to wind load

When pitched roof portal frames are subject to wind or other more irregular loadings, the same basic methods of analysis remain available. The procedure will be illustrated by analysing the frame shown in Fig. 2.61 for a typical case of vertical load together with wind.

Example 2.11 The frame shown in Fig. 2.67 was designed in the previous section for vertical load and found to require the following full plastic moments:

stanchion   311.1 kNm
rafter   207.4 kNm

It is required to check that it is adequate under the factored combined load case shown.

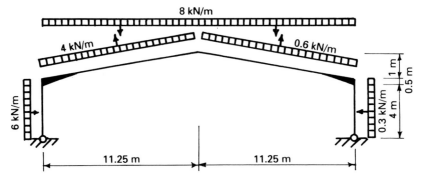

Fig. 2.67 Pitched roof frame subject to vertical and wind load.

The problem will be tackled by applying a load factor λ to the above loads and determining the value of λ at collapse.

The frame is again cut at the apex as shown in Fig. 2.58 so that a free bending moment diagram can be drawn as shown in Fig. 2.68(a) and a reactant bending moment diagram as shown in Fig. 2.68(b).

Combining these two diagrams qualitatively, noting that the pinned bases give rise to a zero moment, gives an impression of the likely shape of the resulting bending moment diagram as shown in Fig. 2.69. It is

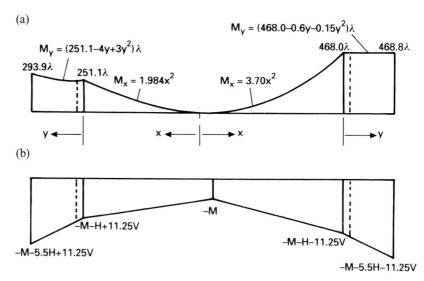

Fig. 2.68 Free and reactant bending moment diagrams: (a) free bending moment diagram; (b) reactant bending moment diagram.

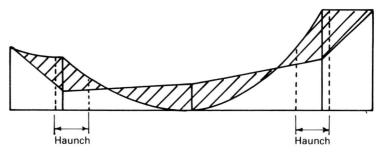

Fig. 2.69 General shape of bending moment diagram at collapse.

immediately obvious that the most likely positions for plastic hinges are near the top of the right-hand rafter and below the right-hand haunch.

If we assume that the rafter hinge forms at the first purlin point at a distance of (say) 1.5 m from the apex, then the following equations can be written down for the bending moments at the pinned bases and the plastic hinge positions:

(left-hand base)	$293.9\lambda - M - 5.5H + 11.25V =$	0
(rafter hinge)	$8.3\lambda - M - H/7.5 - 1.5V =$	$-207.4\,\text{kNm}$
(stanchion hinge)	$467.7\lambda - M - 1.5H - 11.25V =$	$311.1\,\text{kNm}$
(right-hand base)	$468.3\lambda - M - 5.5H - 11.25V =$	0

These four simultaneous equations can be solved to give

$$\lambda = 1.629$$
$$M = 191.6\,\text{kNm}$$
$$H = 78.02\,\text{kN}$$
$$V = 12.62\,\text{kN}$$

The final bending moment diagram can then be drawn as shown in Fig. 2.70 and the solution obtained is thus confirmed by the uniqueness theorem. The value of λ obtained is much greater than unity so that the load case considered is not critical.

Note: The particular version of the semi-graphical method presented here is equally applicable to pinned or fixed base frames and has the advantage of generality. For pinned based frames, which have only one degree of redundancy, simpler solutions can often be obtained by making the frame statically determinate by releasing the horizontal restraint at one base. The horizontal shear at this base then becomes the single redundant force in the solution. This alternative method will be introduced later. However, the method presented here is of more general application and becomes essential when multi-bay frames are considered in Chapter 8.

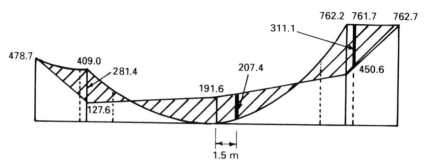

Fig. 2.70 Bending moment diagram at collapse (kNm).

2.6.4   Geometry of the reactant diagram

In many instances, it is useful to be able to adjust the reactant line in order to accommodate such practical factors as discrete member sizes and in order to investigate the effect of, for instance, varying the haunch dimensions. It is therefore often useful to be able to work on an entirely graphical basis. In such cases, the geometric constraints on the reactant diagram can be reduced to two simple rules.

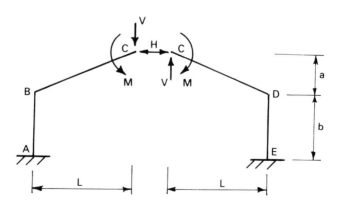

Fig. 2.71 Forces released by cutting the frame.

Consider the general case shown in Fig. 2.71. In order to draw the reactant diagram we need to know the values of the bending moment at the five sections A, B, C, D, E. The equations for these moments are:

$$M_A = -M - (a + b)H + LV$$

$$M_B = -M - aH + LV$$

$$M_C = -M$$

$$M_D = -M - aH - LV$$

$$M_E = -M - (a + b)H - LV$$

These five equations include three unknown forces, M, H and V and eliminating these unknowns gives two equilibrium conditions which the reactant bending moment diagram must always satisfy, namely:

$$M_B - M_A = M_D - M_E$$

$$\frac{M_C - 0.5(M_B + M_D)}{M_B - M_A} = \frac{a}{b}$$

These two conditions are easy to satisfy geometrically, as shown in Fig. 2.72. The first requires that the two dimensions 'y' are equal. The second requires that the dimension 'x' is given by

$$\frac{x}{y} = \frac{a}{b}$$

If the free bending moment diagram is drawn to scale, it is relatively simple to try various reactant diagrams that satisfy these two necessary and sufficient conditions.

The fully graphical method will now be illustrated by means of an example.

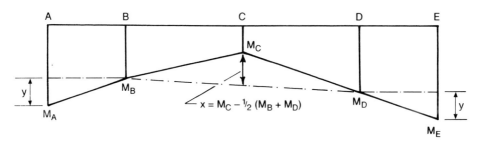

Fig. 2.72 Geometric constraints on the reactant diagram.

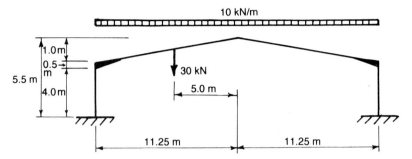

Fig. 2.73 Frame to be designed.

Example 2.12 The frame used in Example 2.9 is to be redesigned:
(a) with fixed bases
and
(b) incorporating a point load from a runway beam as shown for factored loads in Fig. 2.73. It is required to choose suitable sections for the rafters and stanchions.

The free bending moment follows from Fig. 2.65(a) modified for the additional point load. Because the diagram is no longer symmetrical it is necessary to draw it in full and this is done, to scale, in Fig. 2.74.

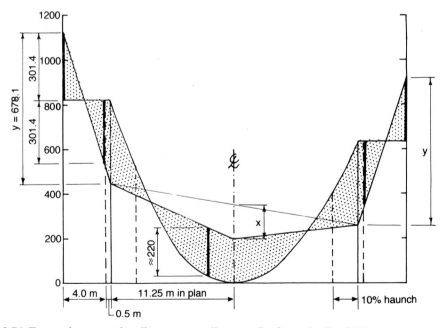

Fig. 2.74 Free and reactant bending moment diagrams for frame in Fig. 2.73.

Figure 2.74 also shows a reactant bending moment diagram and therefore constitutes a solution which is obtained as follows.

It is necessary first to draw the two sloping lines of the reactant diagram for the columns and in order to do this it is necessary to guess a section size for these members. Here we guess a $457 \times 152 \times 52$ kg UB with a full plastic moment of $1096 \times 0.275 = 301.4$ kNm. Recalling that we merely need to draw any reactant line that satisfies the three conditions of equilibrium, mechanism and yield and that, here, the equilibrium conditions are completely satisfied by the two simple rules developed in Section 2.6.4, we assume plastic hinges in both stanchions at the base and below the haunch. This automatically gives equal end slopes with

$$y = 2 \times 301.4 \times \frac{4.5}{4.0} = 678.1 \text{ kNm}$$

It then follows that

$$x = \frac{ay}{b} = \frac{1.0 \times 678.1}{4.5} = 150.7 \text{ kNm}$$

The remainder of the reactant line can now be added as shown on Fig. 2.74.

The full bending moment diagram for the frame is shown shaded and the required full plastic moment for the rafter follows by scaling the maximum bending moment between the haunches. This is approximately 220 kNm so that the required plastic section modulus is given by $220/0.275 = 800 \text{ cm}^3$. A suitable section is therefore a $406 \times 140 \times 46$ UB ($S_{xx} = 889 \text{ cm}^3$). The length of the haunch can also be adjusted at this stage to suit the members chosen.

An experienced designer would immediately suspect that the above design may not be optimum and would question whether a better design may be obtained by increasing the column size and reducing the rafter. If the columns are increased to $457 \times 152 \times 60$ UB with $M_p = 1283 \times 0.275 = 352.8$ kNm, the alternative bending moment diagram shown in Fig. 2.75 follows with

$$y = 2 \times 352.8 \times \frac{4.5}{4.0} = 793.8 \text{ kNm}$$

and

$$x = \frac{1.0 \times 793.8}{4.5} = 176.4 \text{ kNm}$$

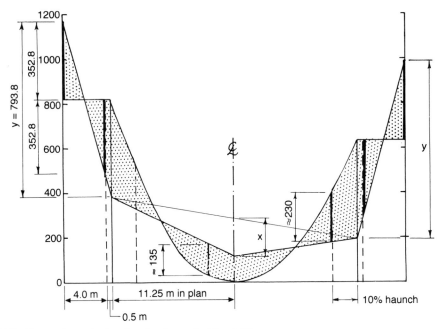

Fig. 2.75 Alternative solution for the frame in Fig. 2.73.

The maximum sagging bending moment in the rafter is approximately 135 kNm requiring a member with a plastic section modulus of $135/0.275 = 491$ cm^3. A suitable section is therefore a $356 \times 127 \times 33$ UB ($S_{xx} = 539$ cm^3). The right-hand part of the bending moment diagram indicates that the length of the haunch would have to be increased beyond the 10% length (2.25 m) shown in order to accommodate this combination of sections. The required length is easily obtained by scaling from the diagram.

It should be noted that having chosen tentative member sizes on the above basis there are a number of stability checks that must be completed before the design can be confirmed. These are considered later.

2.7   Alternative graphical method for pinned-based portal frames

Single-bay portal frames with pinned bases are only one degree redundant and this permits a particularly simple graphical method to be used. This is based on the reasoning shown in Fig. 2.76 in which the frame is made statically determinate by releasing the horizontal force at the right-hand support.

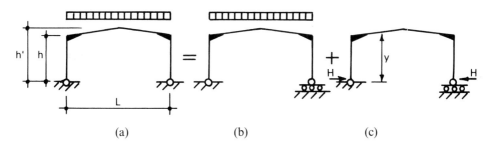

Fig. 2.76 Basis of the alternative graphical method: (a) actual frame and loading = (b) free (with actual loads) + (c) reactant.

Because the structure is only one degree redundant, the reactant diagram involves the single unknown H, the horizontal thrust at the base, and it is convenient to work in terms of this quantity. The reactant bending moment diagram has the magnitude Hy at any point on the frame where y is the height of that point above the base.

The free bending moment diagram is, of course, statically determinate and it is not difficult to construct it, even for quite complex load cases. The two bending moment diagrams can be superimposed on a single diagram as shown is Fig. 2.77 where the shape of the frame is 'unwrapped' for simplicity.

We will therefore illustrate the use of this method by repeating the analysis of the portal frame subject to wind load that was considered as Example 2.11. This method will also be used for the comprehensive design example in Chapter 5 and some frames with other shapes and loading conditions in Chapter 7.

Example 2.13 Figure 2.78 shows a pinned-based frame subject to factored vertical and wind loads. Draw the bending moment diagram at collapse.

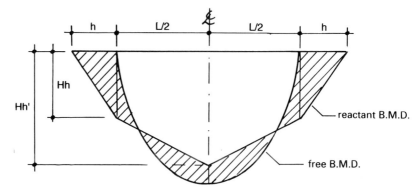

Fig. 2.77 Bending moment diagram for the graphical method.

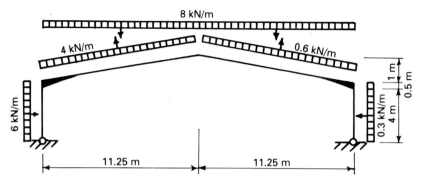

Fig. 2.78 Pitched roof portal frame subject to wind and vertical load.

The first step in the analysis is to find the values of the reactions and the easiest way to do this is to replace the distributed loads by equivalent horizontal and vertical point loads as shown in Fig. 2.79.

The free bending moment diagram can now be drawn. Thus, between A and B with x measured from A to B

$$M_{AB} = 22.25x - \frac{6x^2}{2}$$

From B to C with x measured from B to C in plan

$$M_{BC} = 52.75x - 27\left(2.25 + \frac{x}{11.25}\right) + 22.25\left(4.5 + \frac{x}{11.25}\right)^2$$

$$- (8-4)\frac{x^2}{2} + \frac{4}{2}\left(\frac{x}{11.25}\right)$$

$$= 39.37 + 52.33x - 1.984x^2$$

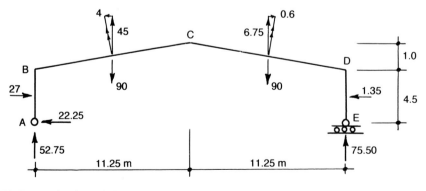

Fig. 2.79 Determination of the reactions at A and E (loads in kN).

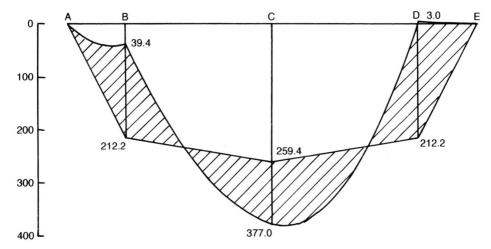

Fig. 2.80 Bending moment diagram for the pinned-based frame (kNm).

From E to D with x measured from E to D

$$M_{ED} = \frac{0.3x^2}{2}$$

and from D to C with x measured from D to C in plan

$$M_{DC} = 75.5x - 1.35\left(2.25 + \frac{x}{11.25}\right) - (8 - 0.6)\frac{x^2}{2}$$

$$+ \frac{0.6}{2}\left(\frac{x}{11.25}\right)^2$$

$$= -3.04 + 75.38x - 3.698x^2$$

The resulting free bending moment diagram is shown drawn to scale in Fig. 2.80. This can be combined with a reactant diagram drawn for any convenient value of the horizontal force H. Here it is drawn for $H = 47.16\,\text{kN}$. This value is not arbitrary, it is in fact the value implicit in the solution to Example 2.11 divided by the load factor of 1.629. It is then easy to see that the shaded bending moment diagram in Fig. 2.80 is the same as that in Fig. 2.70 scaled by 1.629.

2.8   Optimum plastic design

Example 2.12 illustrated that there is often more than one feasible design and it is of interest to enquire whether it is possible to optimise the design in any way. In this section, a procedure is described that is

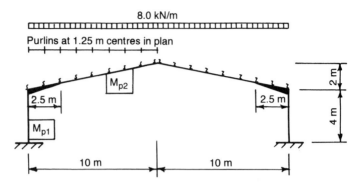

Fig. 2.81 Frame to illustrate minimum weight design.

available whenever there are just two independent member sizes to be chosen. Once again it is convenient to illustrate the method with reference to an example while noting that it is of more general application.

Example 2.14 The fixed based pitched roof structure shown in Fig. 2.81 is to be designed to carry the factored loads shown. Choose optimum values for the full plastic moment M_{p1} of the stanchions and M_{p2} of the rafters.

If we are to attempt to optimise the design, the first question to be addressed is that of the criterion for optimality. The cost of a steel frame is a complex interaction between the cost of basic material together with such matters as fabrication and erection. These latter factors tend to be imponderable outside a particular fabricator's design office. For the purpose of this exercise, therefore, it will be assumed that it will be sufficient to minimise the weight of steel in the members of the frame.

In order to proceed, in a reasonably simple fashion, it is necessary also to assume that there is a continuous spectrum of section sizes available and that, within the members of a particular frame, the relationship between weight per unit length and plastic section modulus or full plastic moment is approximately linear. The validity of these assumptions is illustrated in Fig. 2.82, where the available Universal Beam sections, over a wide range of sizes, are plotted on a graph of area of cross-section versus plastic section modulus.

It can be seen that the mean curves are approximately linear over much of the range shown. However, there is a considerable scatter and, more seriously, the cross-sectional areas of the various sections change in quite large steps which can be as much as 10% of the area. It is

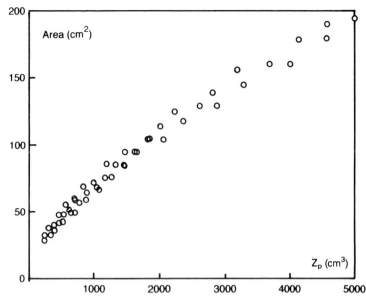

Fig. 2.82 Section properties of Universal Beams.

necessary therefore, to recognise that much of the economic advantage of minimum weight design is likely to be lost when the resulting idealised design is translated into available sections. However, as will be shown later, minimum weight design is not difficult to program for a computer and it then has the additional merit of providing a rational method of automatic design. This aspect is at least as important as the attempt to minimise material consumption.

If the equation of the line relating weight to full plastic moment has the form

$$\text{Weight} = a + bM_p$$

where a and b are constants and if the lengths of member associated with full plastic moments M_{p1} and M_{p2} are L_1 and L_2 respectively, then the weight of steel in the frame, W_f, is given by

$$W_f = L_1(a + bM_{p1}) + L_2(a + bM_{p2})$$

$$= a(L_1 + L_2) + b(L_1 M_{p1} + L_2 M_{p2})$$

It follows that, regardless of the actual values of a and b, the structure of least weight can be found by minimising the weight function

$$Z = L_1 M_{p1} + L_2 M_{p2}$$

In the case of the structure shown in Fig. 2.81, ignoring the length of the haunches, $L_1 = 8$ m and $L_2 = 15.3$ m so that

$$Z = 8M_{p1} + 15.3M_{p2}$$

We now consider the two alternative collapse mechanisms shown in Fig. 2.83. In Fig. 2.83(a), for simplicity, the effect of the finite haunch depth on the position of the plastic hinge in the stanchion is ignored. The work equation for this mechanism is

$$4x^2\phi + 8x(10 - x)\phi = M_{p1}(2\theta + \phi) + M_{p2}\phi$$

$$\text{with} \quad \theta = \frac{x\phi}{20}$$

$$\text{giving} \quad (10 + x)M_{p1} + 10M_{p2} = 40x(20 - x) \tag{2.1}$$

As the plastic hinge in the rafter must fall below a purlin, this equation is valid for $x = 10, 8.75, 7.5, \ldots$, etc.

The alternative mechanism shown in Fig. 2.83(b) is a little more complicated. By considering the two possible expressions for the height of the instantaneous centre of rotation above the plastic hinge at the end of the haunch, we obtain

$$\frac{4.5y}{2.5} = \frac{x + y}{5} \quad \text{giving } x = 8y$$

By equating the two expressions for the horizontal movement of this hinge, we obtain

$$4.5\theta = \frac{9y}{5}\phi \quad \text{giving } \phi = \frac{2y}{5}\phi$$

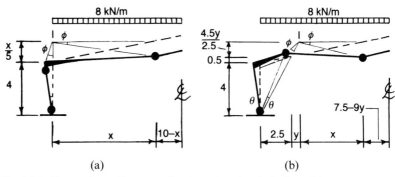

(a)   (b)

Fig. 2.83 Alternative collapse mechanisms for the pitched roof frame.

The work equation can now be written down as

$$8\left(-2.5\frac{y\phi}{2} + 9y \times 3.5y\phi + (7.5 - 9y)8y\phi\right)$$

$$= M_{p1}\frac{2y}{5}\phi + M_{p2}\left(2 + \frac{2y}{5}\right)\phi$$

giving $\quad yM_{p1} + (5 + y)M_{p2} = 100y(11.75 - 8.1y)$ \hfill (2.2)

which is valid for $9y = 7.5, 6.25, 5, \ldots$, etc.

There is also a feasible beam mechanism, which is not shown as it turns out to be not critical. This gives

$$M_{p2} = 28.125\,\text{kNm}$$

By the kinematic theorem (Section 2.1) each of the above mechanisms, with the various possible positions of the plastic hinge near the apex, gives a lower bound on the required values of M_{p1} and M_{p2}. If we plot the various possible equations on a graph of M_{p1} versus M_{p2} we obtain Fig. 2.84. The designations alongside the various lines indicate the relevant equation. Thus, 1a is equation 2.1 with $x = 10$, 1b is equation 2.1 with $x = 8.75$, etc. As we move out from the origin in any direction, the last line encountered must give rise to a family of safe designs and the shaded region of Fig. 2.84 contains all of the permissible designs. Economical designs must evidently lie on the boundary of this feasible region. It now remains to decide which point on this boundary gives the design of minimum weight.

It has already been shown that the weight function to be minimised is given by

$$Z = L_1 M_{p1} - L_2 M_{p2} = 8M_{p1} + 15.3 M_{p2}$$

which is the equation of a straight line on Fig. 2.84. Consider now the arbitrary value of $Z = 2000$ which is shown as a chain-dotted line on Fig. 2.84. All designs lying on this line have equal weight. However, no part of this line is within the feasible region so that all of these designs are unsafe and it is necessary to increase the weight of the structure by choosing a larger value of Z. As Z varies, a series of alternative weight lines are generated which are mutually parallel. Therefore, having drawn a typical weight line, it is simply necessary to move out from the origin parallel to this line until the weight line just touches the permissible region in order to obtain the minimum weight structure which will carry the design loads. The result of this process is the 'tangent weight line' shown in

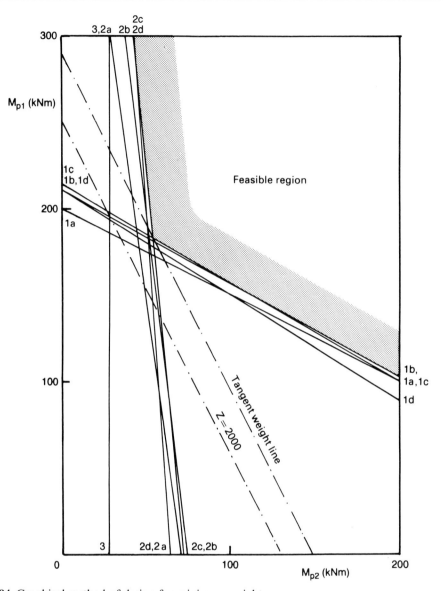

Fig. 2.84 Graphical method of design for minimum weight.

Fig. 2.84 which meets the feasible region at the intersection of lines 1c and 2c. The equations of these lines reduce to

(1c) $1.75M_{p1} + M_{p2} = 375\,\text{kNm}$

(2c) $1.75M_{p1} + 17.5M_{p2} = 1268.8\,\text{kNm}$

so that the minimum weight frame has

$$M_{p1} = 183.3 \, \text{kNm}$$

$$M_{p2} = 54.2 \, \text{kNm}$$

The geometry of Fig. 2.84 indicates that the ratio of M_{p1} to M_{p2} for minimum weight is quite critical. Alternative practical designs lying along line 1c to the right of the critical vertex become increasingly heavy as the ratio of M_{p2} to M_{p1} is increased.

The graphical method of minimum weight design illustrated by Example 2.14 can be used for any structure where the number of independent values of full plastic moment to be chosen is precisely two. The solution will always lie at a vertex of the permissible region where two alternative collapse mechanisms exist simultaneously. For the more general case, where more than two full plastic moments are to be chosen, recourse must be made to mathematical programming as described in the next section.

2.8.1   *General method of minimum weight design*

The solution given above for Example 2.14 can be expressed in more formal mathematical terms. The fact that each of the work equations provided a lower bound to the values of M_{p1} and M_{p2} can be expressed mathematically by writing the equations in the form of inequalities. The problem then becomes:

$$\text{Minimise } Z = 8M_{p1} + 15.3M_{p2}$$

subject to:

(1a) $\qquad 20M_{p1} + 10M_{p2} \geq 4000$

(1b) $\qquad 18.75M_{p1} + 10M_{p2} \geq 3937.5$

- -

(2a) $\quad 0.833M_{p1} + 5.83M_{p2} \geq 416.7$

- -

etc.

This is a formal statement of a typical 'linear programming' problem and the graphical solution given to Example 2.14 is an application of the well-known graphical method of solution for such problems when there are only two variables. However, linear programming problems are particularly amenable to computer solution and standard software is available for the solution of problems with many variables and constraint equations.

Commercial packages for computer-aided plastic design generally use linear programming with automatic equation generation in order to arrive at their initial designs. These designs are then modified in the light of the available range of discrete section sizes and (possibly) also stability requirements.

Chapter 3
Further Considerations in Plastic Design

3.1 Elastic-plastic analysis

Nowadays, more and more structural analysis is being carried out by computer and plastic analysis is no exception. The foundations of the available theories for elastic-plastic analysis were laid in the 1960s and are outside the scope of this book. However, the arrival of the personal computer together with the requirements of modern codes of practice has led to renewed interest in elastic-plastic structural analysis, and commercial packages of use to the practising designer are now appearing on the market. At the very least, the reader of this book needs to know in general terms the sort of product that is on offer.

An elastic-plastic structural analysis usually commences with a conventional elastic analysis using the matrix displacement method. This gives the complete pattern of internal forces (axial force, shear force and bending moment) and displacements under each specified load combination. It then considers each load combination in turn, under increasing load factor, and traces the formation of plastic hinges from first yield to collapse. In the practical packages available to designers, plastic hinges are assumed to form instantaneously when the bending moment at the cross-section reaches its full plastic value (i.e. the members are assumed to have unit shape factor, see Section 1.3). The load-deflection curve is made up of a series of straight lines between plastic hinge formations and the analysis is terminated when the load-deflection curve for some part of the structure becomes horizontal. At this stage, the structure has zero resistance and this implies that a plastic collapse mechanism has been formed.

The load-deflection curve arising from a typical elastic-plastic analysis is shown in Fig. 3.1. As each plastic hinge forms, the designer has available a complete set of information regarding the state of the structure including bending moments, axial forces, shear forces and

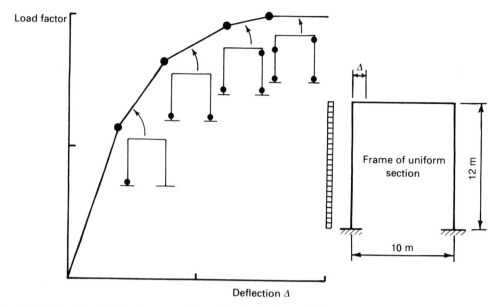

Fig. 3.1 Typical load-deflection curve from elastic-plastic analysis.

deflections. Information regarding intermediate points on the load deflection curve can be obtained by linear interpolation.

The detail arising from such an analysis is, strictly speaking, unnecessary for practical design because the design can be based solely on the bending moment diagram at collapse and the deflections arising from an elastic analysis. Bearing in mind that, with a suitably programmed personal computer, the full analysis can be obtained with a minimum of effort, the advantages of a full elastic-plastic analysis are as follows.

(1) With a complex frame, establishing the correct collapse mechanism by hand is not always easy and errors are not unknown. An elastic-plastic analysis proceeds directly to the critical mechanism.

(2) In the frequently arising circumstances of partial collapse (see Section 2.4), the bending moment diagram at collapse is obtained directly without guesswork or approximation.

(3) The plastic hinge history prior to collapse is known and can be used beneficially. In general, the choice of discrete sections from the available range of Universal Beams, etc. results in a structure that is stronger than is strictly necessary in order to satisfy the design requirements at the ultimate limit state. Elastic-plastic analysis allows the state of the structure to be determined at the specified design load. This means that only plastic hinges present

at this load need to be provided with torsional restraint and that member stability checks can be carried out more precisely. Both of these factors can be used advantageously by the designer.

A further advantage of elastic-plastic computer analysis is that second-order effects can be incorporated relatively easily. The advantage of this will become apparent in the next section.

3.1.1 False mechanisms

Even a brief account of elastic-plastic analysis would be incomplete without mention of the 'false mechanism'. Many designers of steel frames are unaware of this phenomenon and not all the software on the market has facilities for recognising and correcting false mechanisms.

A typical manifestation of the false mechanism arises in the elastic-plastic analysis of a symmetrical pitched roof portal frame subject to uniformly distributed loading. Plastic hinges form in symmetrical pairs and, when either of the hinge arrangements shown in Fig. 3.2 arises, the computer program halts. Sufficient plastic hinges have been formed to constitute a mechanism and the structure has lost its stiffness.

However, the mechanisms shown in Fig. 3.2 are false because the direction of rotation of one of the plastic hinges in the mechanism is in the opposite sense to the bending moment causing the hinge. Sway movement cannot take place without one of the hinges unloading and becoming locked. A robust computer program should be capable of

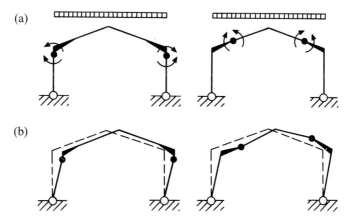

Fig. 3.2 False mechanisms in pinned base portal frames: (a) symmetrical hinge pairs showing direction of bending moment; (b) antisymmetrical sway mechanisms detected by computer. (Plastic hinges shown thus: ●).

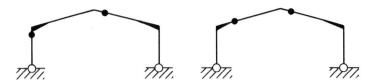

Fig. 3.3 Correct complete collapse mechanisms.

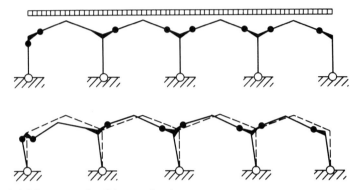

Fig. 3.4 More complex false mechanism.

identifying that these mechanisms are false and of finding the correct 'complete' mechanisms shown in Fig. 3.3.

Figure 3.2 shows two relatively simple manifestations of the false mechanism whose correct collapse mechanisms are shown in Fig. 3.3. There are numerous others and, when multibay portal frames are considered, an almost infinite variety of false mechanisms may arise – Fig. 3.4 shows a typical example. It is therefore of vital importance that engineers using elastic-plastic computer analyses are aware of this phenomenon and that the designers of relevant software include appropriate measures to overcome it.

3.1.2 Transient plastic hinges

Another phenomenon in elastic-plastic analysis which is related to the false mechanism but which requires separate treatment is the transient or 'unloading' plastic hinge. It is possible that, at some stage in an analysis, a plastic hinge tries to reverse its direction of rotation. When this happens, it is necessary to make provision for this hinge to revert to its elastic state while retaining its locked-in plastic rotation. Computationally, this is not trivial but computer programs which do not make provision for this are likely to give wrong answers.

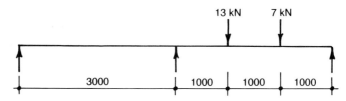

Fig. 3.5 Two-span beam with transient plastic hinge.

The following simple example shows how an elastic-plastic computer program should treat a transient plastic hinge and provides a useful benchmark for checking such programs. It also provides a further illustration of a false mechanism.

Example 3.1 Trace the plastic hinge history of the two-span beam shown in Fig. 3.5 if the full plastic moment of the members is 6.75 kNm and their second moment of area is 400 000 mm^4.

The analysis for the elastic bending moments can easily be done by hand and the resulting bending moment diagram is shown in Fig. 3.6(a). The first plastic hinge forms under the 13 kN load at a load factor of 0.893.

Similarly, the analysis for the bending moment diagram at collapse is trivial using one of the methods given in Chapter 2. As shown in

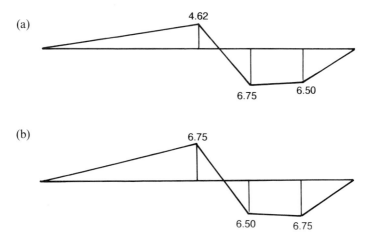

Fig. 3.6 Bending moment diagrams (kNm) for the two span beam: (a) bending moment diagram at first plastic hinge ($\lambda = 0.893$); (b) bending moment diagram at collapse ($\lambda = 1.0$).

Fig. 3.6(b), collapse takes place at a load factor of 1.0 with plastic hinges over the support and below the 7 kN load.

Evidently, during the elastic-plastic phase, the hinge below the 13 kN load must first form and then unload. Figure 3.7 shows the load-deflection curve produced by an elastic-plastic analysis. The first hinge forms at a load factor λ equal to 0.893 and this hinge undergoes plastic rotation until the next plastic hinge forms under the 7 kN load at $\lambda = 0.964$. At this stage, the determinant of the stiffness matrix is zero indicating the existence of a collapse mechanism. However, testing this mechanism indicates that it is a false mechanism with the first hinge to form rotating in the opposite sense to its bending moment. This hinge is therefore locked with its current rotation (0.000 345 rad) and the analysis continued. The next plastic hinge then forms over the support at a load factor of 1.0 to produce the correct collapse load and deflection.

Unloading plastic hinges do not necessarily coincide with spurious mechanisms and, in order to identify transient hinges, it is necessary to continually check that plastic hinges are rotating in the same sense as their bending moments.

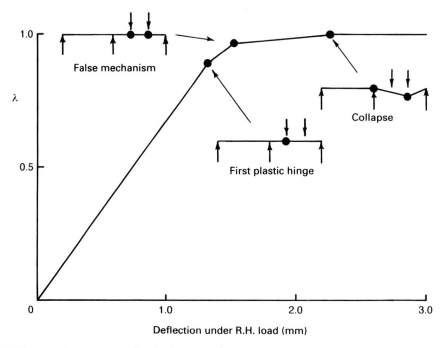

Fig. 3.7 Load-deflection curve for the two-span beam.

3.1.3 *Shakedown*

Although building structures are conventionally designed as though the loading is static, in practice no building ever enjoys static loading. Instead, it is subject to random fluctuations of load throughout its working life. It has been known for many years that when a structure is subjected to variable repeated loading in the elastic-plastic range, repeated applications of the peak loads can cause unlimited plastic flow at load levels below the collapse load of the structure. This plastic flow can take one of two forms.

(1) When a condition of *alternating plasticity* exists in a structure, one or more of the sections of the structure are bent back and forth so that yield of the fibres occurs alternately in tension and compression. Failure by alternating plasticity is therefore a rather severe case of fatigue.

(2) *Incremental collapse* occurs when cyclic application of different combinations of loads causes the progressive development of excessive deflections. When regular cycles of load are applied at a load level above a certain limit, equal increments of load are applied during each cycle of load. These deflection increments are small movements of an *incremental collapse mechanism*. In these circumstances, a few cyclic applications of the peak loads may be sufficient to render the structure useless.

There is a unique limiting load, termed the *shakedown load*, which lies between the yield load and the static collapse load of the structure. Below this load plastic flow is limited but, above this load, cyclic loading can cause one or other of the above forms of unlimited plastic flow.

Shakedown and incremental collapse were the subject of a great deal of research in the formative days of plastic theory and a general method of calculating the shakedown load of a structure was discovered[3.1]. Evidently, this could be used instead of the plastic collapse load in order to define the ultimate limit state of the structure. However, this would be to the considerable disadvantage of plastic theory because not only is some of the economic advantage lost thereby but, more seriously, the shakedown load is much more difficult to calculate.

Fortunately, shakedown analysis is not usually considered to be necessary in building structures and the phenomena of alternating plasticity and incremental collapse have been quietly forgotten. The main reason for this is that it has been demonstrated by an argument based on probability theory[3.2] that a structure is *more likely* to collapse as the result of a single large overload than as the result of the combined effect of several somewhat smaller overloads.

Furthermore[3.3,3.4], it has proved to be difficult to devise practical loading sequences for building structures which lower the shakedown load significantly below the plastic collapse load, and it has been found that strain hardening is particularly beneficial in restricting the continuing movement of incremental collapse mechanisms. However, this latter point should be qualified by the observation that, in structures where frame instability dominates over strain hardening, the reverse is true. A single cycle of overload can weaken the structure with regard to subsequent overloads and cyclic loading can cause 'acceleration to collapse'.

It is probably for this reason that BS 5950: Part 1, in clause 5.7.3.3(c), states that for multi-storey sway frames:

> 'Under all combinations of *unfactored* loading (including the notional horizontal loads when wind loads are not included in the combination) it should be possible by means of moment redistribution to produce sets of moments and forces throughout the frame which are in equilibrium with the applied loads and under which all members remain elastic.'

This clause effectively states that the shakedown load should be considered to be a *serviceability limit* so that a multi-storey structure should not be subject to incremental collapse or alternating plasticity under the working loads. For the reasons given above, this is not an onerous requirement and it is unlikely to influence the design. Nevertheless, it does not seem entirely logical to require this check under the 'simple check for frame stability' and not in the general clause.

Shakedown theory has therefore tended to vanish from the plastic theory syllabus and designers can ignore variable repeated loading with confidence. It does, however, still retain one feature which may prove useful. When using computer-aided design, the disadvantages of more complicated calculation procedures assume less importance. When there are several alternative load cases of roughly equal importance, automatic plastic design becomes problematical. Minimum weight design for the shakedown load[3.5] provides a neat solution to this problem.

3.2   Second-order effects

Much of the important research into second-order effects dates back to the 1950s and 1960s and only relatively minor refinements have been added since then. The subject will be introduced in general terms and can be visualised in terms of (say) a low-rise rectangular framed structure. The important special case of the pitched roof portal frame will be considered in more detail later.

Figure 3.8 shows six idealised load-deflection curves for a typical framed structure. Each of these curves is important and can be obtained analytically. Their particular characteristics are as follows.

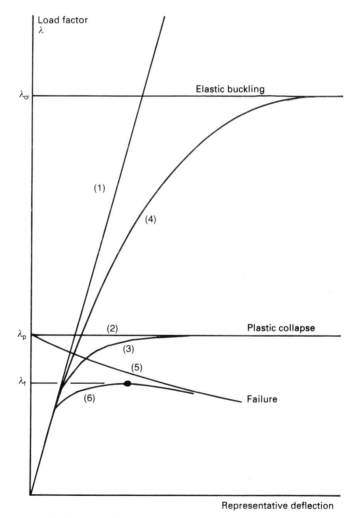

Fig. 3.8 Idealised load-deflection curves for framed structures. Key to curves: (1) Linear elastic; (2) Rigid plastic; (3) Linear elastic-plastic; (4) Second-order elastic; (5) Second-order rigid-plastic; (6) Second-order elastic-plastic.

(1)   *Linear elastic*. This curve is obtained in any conventional elastic analysis which ignores both yield of the material and second-order effects.

(2)   *Rigid-plastic*. This curve is implied in simple plastic theory in which it is assumed that deformation is confined to 'plastic hinges'. No movement of the structure takes place until sufficient hinges to constitute a mechanism have formed. Thereafter, unlimited deflections occur at constant load.

(3)   *Linear elastic-plastic*. If second-order effects are ignored and the load path of the structure is traced under increasing load allowing plastic hinges to form in turn until a complete mechanism is obtained, the result is a linear elastic-plastic load-deflection curve. In all but the simplest of structures, the determination of this curve requires the use of a computer as described in the previous section. The justification for the use of rigid plastic theory in practice lies, of course, in the fact that the same ultimate load is obtained whether rigid-plastic analysis or linear elastic-plastic is used.

Curves (1) to (3) all ignore second-order effects. Curves (4) to (6) are the corresponding curves when second-order effects are involved. They are more difficult to calculate.

(4)   *Second-order elastic*. This curve is obtained when the second-order terms are included in an elastic analysis. The response becomes non-linear from the start and, in the vast majority of structures, axial compressive loads predominate so that it is inherently unstable and elastic failure (without any yielding) takes place at a certain load factor, λ_{cr}.

The predominant second-order effect in steel-framed structures is often termed the 'P–Δ effect' whereby axial compressive loads 'P' in individual members, interacting with finite deflections 'Δ', cause enhanced bending moments and deflections and a consequential loss of stability. This may be visualised as a generalisation of the more familiar behaviour of an axially compressed strut.

Strictly speaking, a distinction should be made between the elastic failure load illustrated by curve (4) in Fig. 3.8 and the 'elastic critical load' obtained when all the member loads are applied axially. The difference between the two is almost entirely academic and for convenience the load factor at elastic failure will be termed λ_{cr}.

(5) *Second-order rigid-plastic.* When the effect of member axial loads is included in rigid-plastic analysis by the work equation method, a drooping load-deflection curve results. This does not represent any possible physical behaviour of the structure but, largely because curve (6) is asymptotic to it, curve (5) can be used to obtain useful estimates of the true failure load.

(6) *Second-order elastic-plastic.* This is, of course, the nearest analytical approach to the true behaviour of a framed structure and the highest point on the curve (load-factor λ_f) represents the best estimate of the actual failure load.

   λ_f is exceedingly difficult to calculate by hand. However, as explained in Section 3.1, it can be calculated relatively easily by computer. Until recently, suitable computer programs have been few in number but they are now beginning to appear in increasing numbers on the commercial market.

At first sight, it appears that the load factor at failure, λ_f. will always fall below the load factor λ_p calculated using simple plastic theory and that this therefore constitutes a serious obstacle in the practical application of plastic theory. Fortunately, this is not the whole story because there is a further beneficial second-order effect, namely strain hardening, which ensures that plastic hinges do not rotate at a constant moment but rather have a rising moment-rotation relationship. Practical structures for which plastic design is appropriate are not usually excessively slender and, in such cases, strain hardening is often sufficient to overcome the destabilising effect of axial compressive loads or at least to ensure that the shortfall of λ_f below λ_p is not excessive. However, the above discussion should be sufficient to cause loud alarm bells to ring if ever a slender structure designed by simple plastic theory is encountered.

It follows that there is a need for methods allowing simple estimates of λ_f (which may or may not take account of strain hardening) and these will be considered in the next section. As the word 'simple' in this context is relative, there will always be a use for the 'exact' computer analysis and this is recommended for unusual structures or for slender structures where the difference between λ_f and λ_p is significant.

3.2.1 The Merchant–Rankine formula

The Merchant–Rankine formula provides the most important approximate method of estimating the load factor λ_f at failure. It was first suggested by Merchant on a purely empirical basis and has a form similar to the better known Rankine equation for struts, namely,

$$\frac{1}{\lambda_f} = \frac{1}{\lambda_p} + \frac{1}{\lambda_{cr}} \tag{3.1}$$

i.e.

$$\lambda_f = \frac{\lambda_p}{1 + \dfrac{\lambda_p}{\lambda_{cr}}} \tag{3.2}$$

Much later, Horne[3.6] showed that the Merchant–Rankine formula had a theoretical basis provided that the plastic collapse mechanism and the lowest buckling mode had similar deflected shapes. If these deflected shapes were dissimilar, the Merchant–Rankine formula was likely to prove conservative.

For relatively stocky frames, it is reasonable to make some concessions for the effect of strain hardening as discussed above. Furthermore, even a small amount of help from the cladding is sufficient to compensate for second-order effects in such frames. Consequently, in clause 5.7.3, which is concerned with the plastic design of multi-storey sway frames, BS 5950 Part 1 includes, among other requirements, the following which is based on an augmented Merchant–Rankine approach.

5.7.3.3   Simple check for frame stability
Clause (d)   In clad frames where no account is taken of the stiffening effect of wall panels the following relationship should be satisfied:

(1) $\lambda_{cr} \geq 4.6$

(2)   when $4.6 \leq \lambda_{cr} < 10$: $\lambda_p \geq \dfrac{0.9\lambda_{cr}}{\lambda_{cr} - 1}$

(3)   when $\lambda_{cr} \geq 10$: $\lambda_p \geq 1$

where λ_{cr} is the elastic critical load factor
λ_p is the rigid plastic load factor of the overall frames but should not be less than 1 locally

Clause (e)   In unclad frames or in clad frames where the stiffness of the cladding is taken into account the following relationship should be satisfied:

(1) $\lambda_{cr} \geq 5.75$

(2)   when $5.75 \leq \lambda_{cr} < 20$: $\lambda_p \geq \dfrac{0.95\lambda_{cr}}{\lambda_c - 1}$

(3)   when $\lambda_{cr} \geq 20$: $\lambda_p \geq 1$

where λ_{cr} and λ_p are as in (d).

The Merchant–Rankine equation and its BS 5950 modification are shown, together with some test results for slender model steel frames in Fig. 3.9[3.7].

It is evident that any approach based on the Merchant–Rankine formula requires that a good estimate of the elastic critical load factor λ_{cr} can be made. For multi-storey frames, this is easier than many engineers realise and a number of accurate approximate methods exist. These include the delightfully simple method given in Appendix F2 of BS 5950 Part 1 which is due to Horne[3.8]. The calculation proceeds as follows.

An accurate linear elastic analysis of the frame is required under side loads applied at each floor level which are equal to 1/200 times the factored vertical loads applied at that level as shown in Fig. 3.10.

At each storey level, the sway index ϕ_s is calculated from

$$\phi_s = \frac{\delta_U - \delta_L}{h}$$

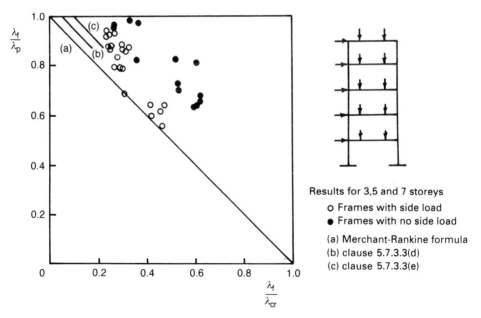

Results for 3,5 and 7 storeys
o Frames with side load
● Frames with no side load

(a) Merchant-Rankine formula
(b) clause 5.7.3.3(d)
(c) clause 5.7.3.3(e)

Fig. 3.9 Merchant–Rankine formula as a lower bound to test results. Results for 3, 5 and 7 storeys: o Frames with side load, ● Frames with no side load. (a) Merchant–Rankine formula, (b) clause 5.7.3.3(d), (c) clause 5.7.3.3(e).

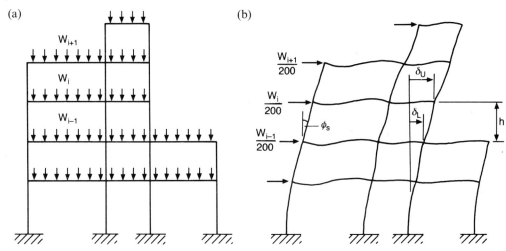

Fig. 3.10 Horne method for calculating λ_{cr} (BS 5950, Part 1, Appendix F2) (a) actual frame; (b) analysis to determine the critical load.

as illustrated in Fig. 3.10. The elastic critical load factor λ_{cr} is given by

$$\lambda_{cr} = \frac{1}{200\phi_{s.max}}$$

where $\phi_{s.max}$ is the largest value of the sway index ϕ_s for any storey of the frame.

Unfortunately, the treatment of second-order effects in pitched roof portal frames is not so simple as will become apparent in the next section.

3.2.2   Second-order effects in pitched roof frames according to BS 5950

Second-order effects in portal frames are covered in BS 5950, Part 1 by clause 5.5.3.2 which accepts three possibilities.

3.2.2.1   A rigorous analysis of frame stability
What is meant by this is not clearly defined but an elastic-plastic second-order analysis leading to the failure load λ_f in Fig. 3.8 is evidently implied. This will give the most accurate assessment of the second-order effects and, in the light of what follows, this, or some other proven method, should be used in preference to the alternatives given in BS 5950 whenever there is any suspicion that second-order effects may be significant. For pitched roof portal frames, the equations given in Sections 3.2.3 to 3.2.8 of this chapter provide a simple solution which is sufficiently accurate for all practical purposes.

3.2.2.2   The fictitious side load method
The paragraph detailing this method reads as follows:

> 'The horizontal deflection δ, calculated by linear elastic analysis, at
> the top of any column due to notional horizontal loading given in
> clause 5.1.2.3 applied in the same direction at the top of each column
> should not exceed $h/1000$ where h is the height of the column. In
> calculating δ, allowance may be made for the restraining effect of
> cladding.'

As with multi-storey frames considered previously, the notional
horizontal forces given in clause 5.1.2.3 are:

> '0.5% of the factored dead load plus vertical imposed load applied
> horizontally.'

The interpretation of the above paragraph is illustrated by Fig. 3.11.
According to BS 5950, second-order effects may be ignored provided
that

$$\delta_1 \text{ and } \delta_2 \le \frac{h_1}{1000}$$

$$\delta_3 \text{ and } \delta_4 \le \frac{h_2}{1000}$$

This treatment appears to be based on two assumptions, namely (a)
the elastic critical load factor λ_{cr} may be estimated with adequate
accuracy by applying the method of Appendix F2 as previously
described in Section 3.2.1, and (b) if $\lambda_{cr} > 5$, there is no need to consider
second-order effects in plastically designed portal frames.

There does not appear to be any definitive study warranting the
different treatment in (b) to the entirely reasonable provisions described
earlier for multi-storey frames. However, assumption (a) is even more
questionable. It presupposes that portal frames buckle elastically in a
sidesway mode due solely to the axial forces in the columns whereas, in

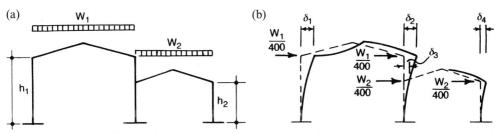

Fig. 3.11 Sway stability check to BS 5950, Part 1, clause 5.5.3.2: (a) factored vertical loads; (b)
notional horizontal loads.

practice, elastic buckling is invariably in a rafter mode as a consequence of the axial compressive force in these relatively longer and more slender members. The critical load for the rafter mode may often be only a small fraction of that for the sidesway mode.

The authors of this clause appear to have compounded their error by allowing designers to include for the restraining effect of cladding in calculating δ. The cladding will generally have a profound effect in resisting sidesway but has little or no effect in resisting a rafter mode. It follows that the sidesway check in BS 5950 gives no useful information regarding the influence of second-order effects in pitched roof portal frames and that it should not be used.

3.2.2.3   The empirical equation method

The relevant paragraph of clause 5.5.3.2 states that irrespective of the effects of cladding, second-order effects may be ignored provided that in each bay of a portal frame

$$\frac{L_b}{D} \leq \frac{44}{\Omega} \frac{L}{h} \left(\frac{\rho}{4 + \rho L_r / L} \right) \left(\frac{275}{p_{yr}} \right)$$

where $\rho = \left(\dfrac{2I_c}{I_r} \right) \left(\dfrac{L}{h} \right)$ for a single-bay frame

   $\rho = \left(\dfrac{I_c}{I_r} \right) \left(\dfrac{L}{h} \right)$ for a multi-bay frame

and L is the span of the bay;

   L_b is the effective span $=$ either L or $L - L_h$ according to the depth of the haunch;

   L_h is the length of the haunch;

   D is the minimum depth of the rafters;

   h is the column height;

   I_c is the minimum second moment of area of the column for bending in the plane of the frame (taken as zero if the column is not rigidly connected to the rafter);

   I_r is the minimum second moment of area of the rafters for bending in the plane of the frame;

   p_{yr} is the design strength of the rafter;

   L_r is the total developed length of the rafter;

   Ω is the arching ratio W_r / W_0;

   W_r is the factored vertical load on the rafters;

   W_0 is the maximum value of W_r which could cause plastic failure of the rafter treated as a fixed ended beam of span L.

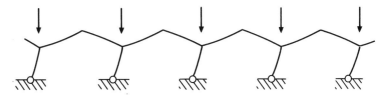

Fig. 3.12 Sway buckling of portal frames.

The derivation of this equation is given in Reference 3.9. It is based on an estimated elastic critical load for sway buckling under the conditions shown in Fig. 3.12 together with the requirement that $\lambda_{cr} > 5$. It would therefore be expected to lead to similar requirements to the fictitious side load method discussed in Section 3.2.2.2 and to be subject to the same criticism. A detailed investigation of this clause[3.10] has verified that this supposition is correct.

3.2.3   Improved treatment of second-order effects in portal frames

It is clear from the foregoing that the elastic critical load λ_{cr} is the crucial parameter in any practical treatment of second-order effects and that the primary problem in BS 5950 is its use of an incorrect estimate of this quantity for pitched roof portal frames. Furthermore, it seems illogical to use different procedures for pitched roof frames and multi-storey frames in the absence of any compelling reason for doing so. For this reason, a more accurate and rational treatment of second-order effects in portal frames has been proposed in two recent papers[3.10,3.11]. The main points made in these papers are summarised here.

The exact elastic critical load can be obtained from the second-order analysis of the model shown in Fig. 3.13. This structure can be analysed using the matrix stiffness method though this generally requires the use

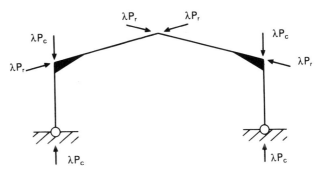

Fig. 3.13 Scheme for elastic critical load calculation.

of a computer. It is not suggested that this is a suitable method for practical design and here it is used as a yardstick whereby alternative methods may be evaluated. It will be shown later that the elastic critical loads calculated on this basis are generally an order of magnitude lower than those predicted by the methods implied in BS 5950.

For a suitable approximate calculation of λ_{cr} for a single-bay pitched roof frame, it is sufficient to consider the half frame shown in Fig. 3.14. The critical mode is assumed to be an anti-symmetrical sway with a corresponding deflection of the rafter. For the present, it is assumed that the values of the axial thrusts P_c and P_r are known. It will be shown later that they can be easily estimated. The important point is that this model includes for the interaction between buckling of the rafter induced by P_r and sway buckling of the column as a result of P_c.

Buckling is initiated by a small disturbing moment M at the eaves which gives rise to a rotation θ. It can be shown that the critical value of λ_{cr}, when the stiffness M/θ becomes zero, is given by the transcendental equation

$$s_r'' + R\left(n_c - \frac{o_c^2}{n_c}\right) = 0$$

where $R = \dfrac{I_c L_r}{I_r h}$

and s'', n and o are stability functions[3.12] with subscripts r and c referring to the rafter and column respectively.

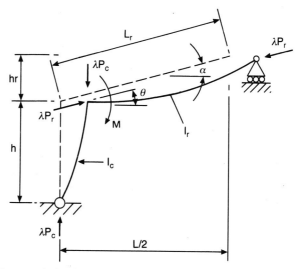

Fig. 3.14 Improved elastic critical load calculation for portal frames.

This equation includes λ_{cr} implicitly in the stability functions and can be solved by trial and error. As it is almost linear in the region of the solution, any reasonable procedure will converge immediately. However, it can be simplified further by introducing approximations for the stability functions to give the following explicit expression for λ_{cr}.

$$\lambda_{cr} = \frac{3EI_r}{L_r\left[\left(1+\dfrac{1.2}{R}\right)P_ch + 0.3P_rL_r\right]}$$

The corresponding transcendental equation for a fixed base frame is

$$s_r'' + Rn_c = 0$$

which simplifies to

$$\lambda_{cr} = \frac{5E(10 + R)}{\dfrac{5P_rL_r^2}{I_r} + \dfrac{2RP_ch^2}{I_c}}$$

The above equations give rise to simple estimates of the elastic critical load which have been shown by parametric study to be close enough to the exact values for all practical purposes.

An estimate of the axial loads P_c and P_r that is also sufficiently accurate for practical design can be obtained as follows, using the notation given in Fig. 3.15. In this figure, M is the eaves moment at collapse which has been found during the plastic design.

$$P_c = \frac{wL}{2} \qquad H = \frac{M}{h}$$

$$P_r = H\cos\theta + V\sin\theta$$

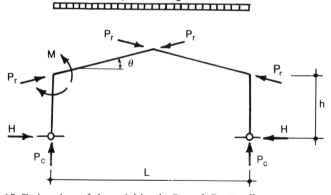

Fig. 3.15 Estimation of the axial loads P_c and P_r at collapse.

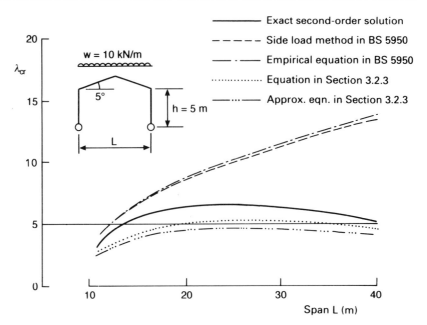

Fig. 3.16 Elastic critical loads for pinned based frames of 5 m height and 5° roof pitch.

A typical comparison of the elastic critical loads given by exact analysis, the above method and the approximate methods implied in BS 5950 is given in Fig. 3.16. It is immediately evident that the proposed method gives an adequate estimate of the critical load whereas BS 5950 is grossly unsafe.

3.2.4   *Influence of partial base fixity*

In clause 5.1.2.4, BS 5950 recognises three types of base stiffness:

(1)   Where the column is rigidly connected to a suitable foundation, the stiffness of the base should be taken as equal to the stiffness of the column.

(2)   Where the column is nominally connected to the foundation, a base stiffness of 10% of the column stiffness may be assumed.

(3)   Where an actual pin or rocker is provided, the base stiffness should be taken as zero.

It is also stated that it is necessary to be consistent in using the same base stiffness in all calculations. This includes:
elastic frame analysis (for deflections);
effective length of columns;

calculation of the elastic critical load;
classification as sway or non-sway.

It follows that BS 5950 does not recognise a fully fixed base and that, in general, it is not necessary to treat a nominally pinned base as having zero stiffness. The wider implications of this clause are discussed in Reference 3.13, here we concentrate on the consequences for the elastic critical load calculation. This is considered in Reference 3.14 as follows.

In a global calculation, a nominally fixed base can be modelled as a spring with rotational stiffness equal to $4EI_{column}/L_{column}$. If the computer program being used cannot accept a rotational spring, the model shown in Fig. 3.17 may be used in which the base stiffness is modelled by a dummy member continuous with the column and pinned at the far end which has a stiffness equal to EI_{column} and a length of $0.75L_{column}$.

Similarly, a nominally pinned base can be modelled as a spring with rotational stiffness equal to $0.4EI_{column}/L_{column}$. If the computer program being used cannot accept a rotational spring, the model shown in Fig. 3.17 may again be used but with the I of the dummy member reduced to $0.1I_{column}$.

It should be noted that the model shown in Fig. 3.17 affects the reaction at the column base which should be corrected to be equal to the axial force in the column. Reference 3.14 also includes improved explicit expressions for the elastic critical load λ_{cr} which, with the notation of Section 3.2.3, are as follows.

Nominally pinned base:

$$\lambda_{cr} = \frac{(4.2 + 0.4R)EI_r}{L_r\left[0.42P_rL_r + \left(1.16 + \dfrac{1.2}{R}\right)P_ch\right]}$$

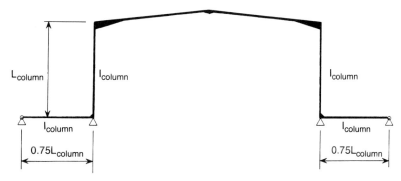

Fig. 3.17 Alternative model for base flexibility.

Nominally fixed base:

$$\lambda_{cr} = \frac{5E(10 + 0.8R)}{\left[\dfrac{5P_R L_r^2}{I_r} + (2.6R + 4)\dfrac{P_c h^2}{I_c}\right]}$$

3.2.5   Stability of multi-span portal frames

It has been shown[3.11] that the simple equations for the calculation of the elastic critical load given in the previous sections are equally applicable to the overall stability of multi-span frames when applied to the outer spans. If the frame is irregular, as typified by Fig. 3.18, the equations should be applied to both outer spans and the lowest critical load so obtained should be taken as being applicable to the complete frame. This procedure will generally be conservative. The internal rafters of multi-span frames should also be checked according to Section 3.2.6.

3.2.6   Stability of the internal rafters of multi-span frames

In BS 5950: Part 1, the stability of internal rafters is considered in clause 5.5.3.3 which is entitled 'Snap through stability of rafter'. This title is not strictly accurate as it is not a 'snap-through' that is critical but buckling of an internal rafter as illustrated by Fig. 3.19. Despite the confusion caused by the title, this clause has been found[3.11] to provide an adequate basis for practical design. It contains a restriction on the rafter slenderness L_b/D where L_b is the effective span of the bay, and D is the minimum depth of the rafters.

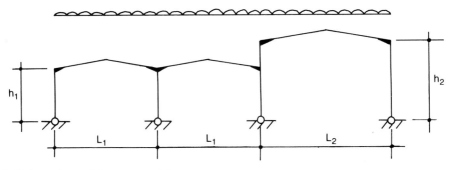

Fig. 3.18 Irregular multi-span portal frame.

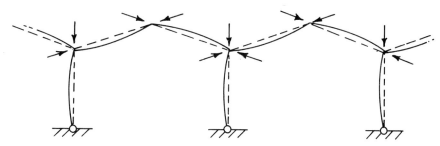

Fig. 3.19 Buckling of the internal rafters of a pitched roof frame.

The background to this clause is given in a paper by Horne[3.9] in which he first derives an approximate expression for the elastic critical load λ_{cr} and then imposes the restriction that $\lambda_{cr} \geq 2.5\Omega$ where Ω is the arching ratio defined below. For practical frames, this can result in values of the critical load between 4 and 10. Within the philosophy upon which this chapter is based, this requirement is best expressed in the form of a calculation of λ_{cr} which can then be used in the same way as other values of the critical load found for other modes. Thus:

$$\lambda_{cr} = \left(\frac{D}{L_b}\right)\left(\frac{55\left(4 + \dfrac{L}{h}\right)}{\Omega - 1}\right)\left(1 + \frac{I_c}{I_r}\right)\left(\frac{275}{p_{yr}}\right)\tan 2\theta_r$$

where θ_r is the slope of the rafter and the remaining symbols have all been defined in Section 3.2.2.3.

It should be noted that it is possible to design frames with very low load factors with respect to this buckling mode so that this check is essential.

3.2.7 Stability of portal frames with internal valley beams or props

Before consideration of the stability of multi-span portal frames is complete, it is necessary to consider a situation which occurs occasionally in practice where some or all of the internal valleys are supported on valley beams or otherwise supported on non-moment-bearing props. The stability of these arrangements, which are illustrated in Fig. 3.20, will be covered by the checks described earlier unless a sidesway mode is critical. However, the reduced sidesway stiffness of these configurations means that they are much more susceptible to sway buckling and this must be checked.

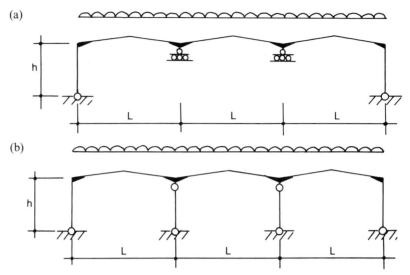

Fig. 3.20 Frames with simply supported internal valleys: (a) frames supported on valley beams; (b) frames with internal props.

It is important to realise that, although the two situations in Fig. 3.20 give rise to similar plastic designs, they require separate consideration with regard to stability. The elastic critical loads will be quite different because the axial force carried by an internal prop will act as a 'destabilising load' during sidesway buckling. Thus, configurations such as the one shown in Fig. 3.20(b) are far less stable than configurations like Fig. 3.20(a) and designers should realise this. An analysis is required which recognises this and this is given in Reference 3.11.

With truly pinned bases, the elastic critical load may be estimated using

$$\lambda_{cr} = \frac{3EI_r}{2L_r\left[\left(1 + \dfrac{1.2}{R_p}\right)(N+1)P_ch + 0.6P_rL_r\right]}$$

where N is the number of internal props ($=0$ for valley beams as in Fig. 3.20(a))

$$R_p = \frac{\text{stiffness of column}}{\text{stiffness of rafter pair}} = \frac{2I_cL_r}{I_rh}$$

and the remaining symbols are as before.

A parametric study of some typical frames of this type has revealed that the elastic critical load may be of the order of unity which indicated that these frames, even though rationally designed and of practical proportions, are potentially very unstable. A check such as the one given above should therefore be regarded as essential. The side load method given in BS 5950; clause 5.5.3.2 is *not* adequate in this respect because, as before, it ignores the destabilising effect of the axial loads in the rafters and it also ignores the equally important destabilising prop loads.

However, the stiffening effect of the cladding (stressed skin design) is particularly effective in resisting this form of instability. This is presumably the reason why there is no history of unexpected collapses of frames supported on valley beams, and stressed skin design may provide a design solution in cases where increasing the size of the members in order to avoid sway instability would be too expensive.

3.2.8   *The design of portal frames taking into account second-order effects*

The in-plane stability considerations in portal frames designed by plastic theory are similar to those for multi-storey frames and it is logical that they should both be treated in the same way. In the absence of a rigorous second-order analysis, the crucial parameter is the elastic critical load λ_{cr}. Simple equations have been given whereby λ_{cr} may be estimated for the following modes of instability:

❏ buckling of outer rafter combined with column sway;
❏ buckling of internal rafter (confusingly termed 'snap through' in BS 5950);
❏ sway of multi-bay frames supported on valley beams or props.

The critical load of the frame is the lowest of these when all relevant modes are considered. It should be noted that the cladding (stressed skin or diaphragm effect) will only be of significant benefit with respect to the third of these.

The value of λ_{cr} found in this way should then be used as follows. This is exactly the same as for multi-storey sway frames in clause 5.7.3 of BS 5950: Part 1.

If $\lambda_{cr} \geq 10$, in plane instability effects may be ignored, i.e. the rigid-plastic load factor may be used without reduction.

If $4.6 \leq \lambda_{cr} \leq 10$, the required load factor for plastic design should be increased to

$$\lambda_p \geq \frac{0.9\lambda_{cr}}{\lambda_{cr} - 1} \quad (= \beta \geq 1)$$

If $\lambda_{cr} \leq 4.6$, in-plane instability effects should be taken into account by means of an elastic-plastic second-order analysis of the complete frame.

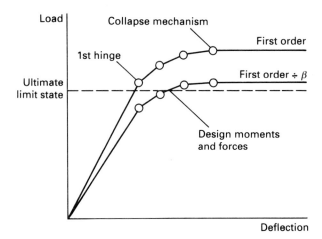

Fig. 3.21 Load-deflection curves showing plastic hinge history.

Consideration of some typical portal frames suggests that well-designed frames of conventional proportions will generally have values of λ_{cr} round about 5 so that the design for second-order effects is very important.

It follows that the majority of practical frames will fall into the intermediate case and that the approximate treatment of second-order effects will be widely used in practice. This requires that the load factor for plastic design should be increased according to the above equation for λ_p and that all of the internal forces obtained by first-order analysis should be amplified in proportion. It is not easy to see how this may be achieved when using plastic theory without exceeding the full plastic moment of the members! Reference 3.14 offers a solution.

For many reasons, modern plastic design is best carried out with the aid of a first-order elastic-plastic analysis which gives rise to a plastic hinge history as shown by the upper curve in Fig. 3.21. Factoring the applied loads by β, giving the lower curve, has the same effect as amplifying the internal forces by β. Where this lower curve crosses the ultimate limit state design loads gives the conditions under which the member stability checks should be carried out. These can easily be extracted from the results of the analysis.

3.3 Member stability

It is essential that all members in a structure remain stable up to the required ultimate limit state loading condition. In structures that are designed to remain elastic at the ultimate limit state, this means ensuring that the members are stable against both in-plane and out of plane failure. In plastic design, the possible presence of plastic hinges at the ultimate limit state places greater emphasis on ensuring that there is not premature out of plane failure of any member.

In structures that are designed using plastic theory, there is the overriding requirement given in BS 5950: Part 1, clause 3.5.2 that only Class 1 (plastic) sections may be used for plastic design. Class 1 sections are defined in Table 7 in terms of limiting width to thickness ratios for the various elements (compression flange outstands, webs, etc.) of a member. These serve to ensure that the element is sufficiently stocky for a plastic hinge to develop and to have adequate rotational capacity without premature loss of strength due to local buckling. Clause 5.3.4 adds a further requirement that plastic hinges are only allowed in members which are symmetrical about an axis perpendicular to the axis of the hinge rotation. There is also the concession that members which do not contain plastic hinges need only be Class 2 (compact).

The next critical requirement, specified in clause 5.3.5, is that torsional restraints should be provided at all plastic hinge locations. Where this is impractical, the restraint should be provided within a distance $D/2$ of the plastic hinge location along the member. There is then the further requirement that there should also be an adjacent restraint within a distance L_m of the hinge restraint. The calculation of L_m is important and will be considered in some detail later. These requirements for torsional restraints are also necessary in order to ensure that plastic hinges have adequate rotational capacity.

These requirements may be relaxed when it can be demonstrated that there is no need for a particular plastic hinge to achieve a significant plastic rotation in order for the structure to reach its design strength. In clause 5.5.3.1, therefore, it is stated that *in a portal frame* it is not necessary to provide torsional restraints at the last hinge to form *provided it can be clearly identified*. In many portal frames, the dominant loading condition is $1.4 \times$ Dead $+ 1.6 \times$ Imposed $+$ Notional Horizontal Loads and this usually results in the last hinge forming as a sagging hinge in the rafter adjacent to the apex. It is generally worthwhile confirming that the rafter hinge is the last hinge to form in order to reduce the number of costly bracing members and to allow a greater purlin spacing along the rafter.

It follows that, in a member containing a plastic hinge, the distance between effective restraints must usually be less than it would be in a similar elastically designed member where there is no requirement for rotational capacity. This can be one of the more troublesome aspects of plastic design.

3.3.1   Overview of the clauses dealing with member stability

In BS 5950: Part 1, there are various formulae of differing degrees of complexity for checking member stability. Usually the more complex formulae give more efficient solutions at the expense of the designer's time. The main justification for the more complex procedures is that the complexity is irrelevant once the formulae are included in computer programs.

An overview of the available procedures for verifying member stability is given in Table 3.1. The three main factors that determine which approach is appropriate in a particular case are:

❏ Whether the member is restrained at intervals along the tension flange. This is an important feature which appeared in a design code for the first time in BS 5950. It is particularly applicable to the stanchions and haunch region of a portal frame where tension flange restraint is provided by the purlins and sheeting rails.

❏ Whether the length of member under consideration contains a plastic hinge. As pointed out above, the requirements for member stability become more onerous in the vicinity of a plastic hinge.

❏ Whether the member is of uniform or non-uniform section. In general, the plastic design of tapered members is not encouraged. However, the haunch region of a portal frame is an important exception and most plastically designed portal frames contain non-uniform members even though the tapered section is itself designed to remain elastic.

The salient features of the three relevant parts of BS 5950: Part 1 are as follows.

Clause 4.8.3 which depends also on clause 4.3.7, is the general clause which is concerned solely with elastic conditions and is only appropriate for checking lateral torsional buckling when the length of member under consideration remains elastic or at most contains only the last hinge to form.

Clause 5.3.5 contains a conservative formula for the spacing of restraints in a member containing a plastic hinge.

Clause 5.5.3.5 contains simplified formulae for checking the stability of portal frame rafters with and without plastic hinges including restraints to the tension flange.

Appendix G contains more elaborate formulae for checking the stability of members that have intermediate restraints along the tension flange. However, the calculations required by Appendix G are far from trivial and in practice it is mainly used when suitable computer software is available.

Table 3.1 Selection table for clauses or formulae for member stability.

			Clause or formula	Notes
Unrestrained length	No plastic hinge present	Uniform member	4.8.3	Standard elastic check
		Non-uniform member	4.8.3	
	Plastic hinge present	Uniform member	5.3.5	Conservative, for constant bending moment
		Non-uniform member	5.3.5	A non-uniform member containing a plastic hinge should be avoided
		Uniform member	Ref. 3.15	Most efficient. Computer/calculator program required
		Alternative formulae not in BS 5950	3.3.3 of this book	Simplified version of Ref. 3.16
Length restrained along the tension flange	No plastic hinge present	Uniform member	G.2(a).(1)	Purlin/rail centres to satisfy 4.8.3.3
		Non-uniform member	G.2(a).(2)	
	Plastic hinge present	Uniform member	G.2(b).(1) or (2)	Purlin/rail centres to satisfy 4.8.3.3 (Not 5.3.5 refer to G.1.2.2)
		Non-uniform member	G.2(b).(2)	
	Simple formulae, with or without a plastic hinge	Uniform and non-uniform members	5.5.3.5	Purlin centres to satisfy 5.3.5 if a plastic hinge exists or 4.8.3.3 if not

The member stability clauses in BS 5950 also make reference to several different length parameters and it is also helpful to identify and define these.

'L' (clause 4.3.5) is the actual length, or portion of the length, of the member under consideration. For a beam, it can be the span or the distance between restraints.

'L_E' (clause 4.3.5) is the *effective* length of a member which is required for the calculation of the bending capacity taking account of lateral torsional buckling. L_E is dependent upon the amount of restraint that the member receives at its ends and also on the loading conditions. Guidance is given in Tables 9 and 10 of BS 5950: Part 1. The procedure for using L_E is as follows.

L_E is used to obtain the slenderness ratio $\lambda = L_E/r_y$. This then gives rise to the equivalent slenderness $\lambda_{LT} = nuv\lambda$. The bending strength p_b is then a function of λ_{LT} and the design strength p_y according to Tables 11 and 12. The bending capacity of the section is then given by $M_b = p_b S_x$ where S_x is the plastic section modulus of the section.

In the above:

n is a slenderness correction factor which is described in clause 4.3.7.6 and which is a function of the loading and restraint conditions on the member. n may be conservatively taken as unity.
u is a buckling parameter which is tabulated for the standard sections. For rolled I or H sections, it may be conservatively taken to be 0.9.
v is a slenderness factor which is a function of λ and the section properties of the member. It may be determined using Table 14 of BS 5950 or conservatively taken to be equal to unity.

'L_m' (clause 5.3.5) is the maximum permissible distance between the torsion restraint at a plastic hinge position and the nearest adjacent torsional restraint. L_m, as given in clause 5.3.5, is conservative and other more effective formulae are given in Section 3.3.3 of this book.
'L_s' (clause 5.5.3.5 and Figure 10) is concerned with portal frame rafters and is the maximum permissible distance between restraints to the *compression* flange when there are restraints at intervals along the *tension* flange. The spacing of the tension flange restraints must satisfy the L_E calculation described above.
'L_t' (Appendix G) is the maximum permissible distance between torsional restraints in members which are restrained at intervals along the tension flange. The calculation procedure in Appendix G is of rather more general application and includes non-uniform (tapered) members.

3.3.2 Restraints

BS 5950 makes reference to two distinct types of restraints although both are often loosely referred to as simply 'restraints'. It is important to distinguish between them and to be clear when one or the other is required.

Torsional restraints are required at and adjacent to plastic hinge positions. They are defined in clause 4.3.3 as follows.

'A beam may be taken as torsionally restrained (about its long-itudinal axis) at any point in its length where *both flanges* are effectively held in position relative to each other by external means (in the lateral direction).'

Clause 4.3.3 also describes how torsional restraint may be achieved.

'*Lateral restraints* inhibit lateral movement of the *compression flange* of a member relative to the supports. The requirements for effective lateral restraint are given in some detail in clause 4.3.2.

Unless a member is continuously restrained (e.g. by a concrete floor slab), lateral restraints are required in order to prevent the lateral torsional buckling of members. It is usually then necessary to check that the unrestrained length of the member is stable between points of lateral restraint.'

As the requirements for torsional restraints are more onerous than those for lateral restraints, torsional restraints also provide lateral restraint (clause 4.3.1).

The lateral restraint to the rafters of a portal frame is usually provided by the purlins and these are in turn restrained by the sheeting. Clause 4.3.2.4 makes the important point that in this circumstance, it is *not* necessary to check the purlins for the induced restraining forces provided that either:

(1) there is bracing of adequate stiffness in the plane of the rafters; or
(2) the roof sheeting is capable of acting as a diaphragm.

Evidently, care is required here with some modern types of roof such as standing seam and clip-fix systems which offer only limited restraint to the purlins.

3.3.3   *Influence of the shape of the bending moment diagram*

Designers having experience of elastic design will know that the shape of the bending moment diagram has a considerable effect on the allowable distance between discrete lateral restraints. In BS 5950: Part 1, the elastic bending moment profile in uniform members is accounted for in the lateral torsional buckling clause 4.3.7 in one of two ways as summarised in Table 13. If the bending moment diagram is linear between adjacent lateral restraints, the maximum bending moment M is reduced by a factor 'm' which in turn depends on the ratio 'β' between the end moments. This procedure is illustrated in Fig. 3.22.

If the bending moment diagram is non-linear between adjacent lateral restraints, there is an alternative procedure which results in a reduction factor 'n' on the slenderness ratio. It is not permitted to use both m and n.

Discounting any slight curvature of the bending moment diagram caused by the self-weight of the member, it is unusual to find bending moment diagrams which are not at least approximately linear between restraints and therefore the equivalent moment factor m based on the end moment ratio β is used far more frequently than the slenderness reduction factor n.

In BS 5950, there is no equivalent procedure based on the shape of the bending moment diagram for members containing plastic hinges. This is surprising since it is clear that a plastic hinge in a region of uniform or approximately uniform bending moment is potentially very unstable as both flanges of the member would be yielding over a significant length. Conversely, a plastic hinge in a region of rapidly changing bending moment is a much less serious matter as the plastic zones would be relatively confined. The relevant clause 5.3.5 is conservative because it determines the maximum restraint spacing L_m for the worst case, namely a constant bending moment. This is the reason that the restraints required by clause 5.3.5 are so close together.

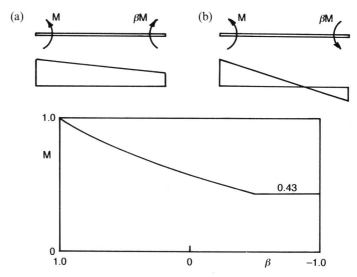

Fig. 3.22   Equivalent uniform moment factor 'm': (a) β positive; (b) β negative.

The much more complicated formulae given in Appendix G do take account of plastic hinges in members with non-uniform bending moment but these are only for members restrained at intervals along the tension flange.

There is evidently a need for a relatively simple method for determining a less conservative value of L_m in members subject to a bending moment gradient. A suitable procedure is described in Reference 3.16 and this is advocated for general use. It is illustrated in Fig. 3.23 where the restraint spacing L'_m is a function of the end moment ratio β defined in Fig. 3.22.

In Fig. 3.23, L_m is determined using the formula given in clause 5.3.5 of BS 5950: Part 1, namely

$$L_m = \frac{38r_y}{\left[\dfrac{f_c}{130} + \left(\dfrac{p_y}{275}\right)^2 \left(\dfrac{x}{36}\right)^2\right]^{\frac{1}{2}}}$$

where $f_c =$ the average compressive stress due to axial load (N/mm^2)
   $p_y =$ the design strength (N/mm^2)
   $r_y =$ the radius of gyration about the minor axis (mm)
   $x =$ the torsional index

This formula holds for values of β between 1 and a limiting value given by

$$\beta_m = 0.44 + \frac{x}{270} - \frac{f_c}{200} \quad \text{for Grade 43 steel}$$

$$\beta_m = 0.47 + \frac{x}{270} - \frac{f_c}{250} \quad \text{for Grade 50 steel}$$

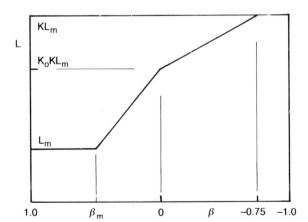

Fig. 3.23 Restraint spacing in a member subject to moment gradient.

For values of β less than this, the restraint spacing can be increased up to a maximum of KL_m at $\beta = -0.75$ where

$$K = 2.3 + 0.03x - \frac{xf_c}{3000} \qquad \text{when } 20 < x \le 30$$

$$K = 0.8 + 0.08x - \frac{(x-10)f_c}{2000} \qquad \text{when } 30 < x \le 50$$

As can be seen in Fig. 3.23, the method provides for a bilinear transition from the point with coordinates (β_m, L_m) to $(-0.75, KL_m)$ via $(0, K_0 KL_m)$, where

$$K_0 = \frac{180 + x}{300}$$

The required restraint spacing L'_m is then:

$$\text{when} \quad 1.0 > \beta > \beta_m \qquad L'_m = L_m$$

$$\beta_m > \beta \ge 0 \qquad L'_m = \left[1.0 + \frac{\beta_m - \beta}{\beta_m}(K_0 K - 1)\right] L_m$$

$$0 > \beta \ge -0.75 \qquad L'_m = K\left[K_0 - \frac{\beta(1 - K_0)}{0.75}\right] L_m$$

$$-0.75 > \beta \qquad L'_m = KL_m$$

The procedure described above is a simplification of a more exact analysis which was derived by Horne[3.16]. It has been verified by comparison with Horne's analysis over the whole range of Class 1 Universal Beam sections, and a typical comparison is shown in Fig. 3.24 for a $533 \times 210 \times 82$ UB with $f_c = 15 \, \text{N/mm}^2$ and $p_y = 275 \, \text{N/mm}^2$.

Table 3.2 is a design aid which can be used to simplify the above calculation. The above method can be used for values of f_c up to $80 \, \text{N/mm}^2$ provided that f_c does not exceed the axial stress that changes the section classification from plastic to compact.

Evidently, there has to be an appreciable moment gradient before the L_m condition in clause 5.3.5 of BS 5950: Part 1 can be relaxed. In the important case of the rafter hinge in a portal frame, the bending moment diagram is nearly constant and no relaxation is possible. This is a further incentive to ensure that the rafter hinge is the last to form so that this region can be designed elastically and the restrictions of clause 5.3.5 avoided.

Further information on this important subject of the lateral stability of steel beams and columns can be found in a Steel Construction

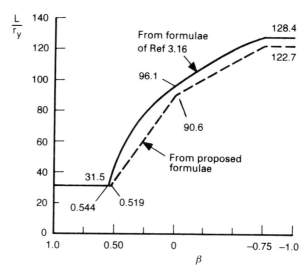

Fig. 3.24 Comparison of exact[3.15] and approximate[3.16] restraint spacings.

Institute publication on the subject[3.17]. This details the common cases of restraint but has very little to say on the important subject of members containing a plastic hinge.

3.3.4   Member stability calculations for beams

3.3.4.1   Beams supporting concrete floors

Beams supporting concrete floors, particularly non-composite precast concrete units, are a common form of construction. Reference 3.17 gives recommendations for checking whether or not the friction between the concrete units and the top flange of the steel beam is sufficient to provide restraint. The precast units need to be adequately grouted, ideally with a top screed, in order to ensure that they function as one effective restraining slab rather than as individual concrete units.

For a simply supported beam of span L carrying an unfactored uniformly distributed load W, the beam may be designed as fully restrained if

$$W\mu > r\gamma_f \frac{WL}{8D}$$

where r = restraint force coefficient (ratio of the required bracing force to the maximum compression flange force in the beam)

Table 3.2 Spacing of torsional restraints according to Section 3.3.3.

(1) $p_y = 275\,\text{N/mm}^2$ $f_c = 0\,\text{N/mm}^2$

x	L_m/r_y	β_m	L'_m/r_y		
			$\beta = 0.25$	$\beta = 0$	$\beta = -0.75$
20	68.4	0.514	101.2	132.2	198.4
25	54.7	0.533	86.2	114.0	166.9
30	45.6	0.551	76.5	102.1	145.9
35	39.1	0.570	73.7	100.8	140.7
40	34.2	0.588	72.2	100.3	136.8
45	30.4	0.607	71.5	100.3	133.8
50	27.4	0.625	71.4	100.7	131.3

(2) $p_y = 275\,\text{N/mm}^2$ $f_c = 15\,\text{N/mm}^2$

x	L_m/r_y	β_m	L'_m/r_y		
			$\beta = 0.25$	$\beta = 0$	$\beta = -0.75$
20	58.4	0.439	80.1	108.9	163.4
25	49.2	0.458	71.4	98.2	143.8
30	42.2	0.476	65.0	90.2	128.8
35	36.9	0.495	63.3	90.2	125.9
40	32.7	0.513	62.4	90.5	123.5
45	29.3	0.532	62.0	91.0	121.4
50	26.6	0.550	62.1	91.7	119.6

(3) $p_y = 275\,\text{N/mm}^2$ $f_c = 30\,\text{N/mm}^2$

x	L_m/r_y	β_m	L'_m/r_y		
			$\beta = 0.25$	$\beta = 0$	$\beta = -0.75$
20	51.7	0.364	64.7	93.1	139.7
25	45.0	0.383	59.2	86.1	126.0
30	39.5	0.401	54.8	80.2	114.6
35	35.0	0.420	53.6	81.0	113.0
40	31.4	0.438	53.0	81.7	111.4
45	28.4	0.457	52.9	82.5	110.0
50	25.9	0.475	53.1	83.3	108.6

Table 3.2 Spacing of torsional restraints according to Section 3.3.3 (continued).

(4) $p_y = 355 \, \text{N/mm}^2$ $f_c = 0 \, \text{N/mm}^2$

x	L_m/r_y	β_m	L'_m/r_y		
			$\beta = 0.25$	$\beta = 0$	$\beta = -0.75$
20	53.0	0.544	79.7	102.4	153.7
25	42.4	0.563	67.9	88.3	129.3
30	35.3	0.581	60.3	79.1	113.0
35	30.3	0.600	58.2	78.1	109.0
40	26.5	0.618	57.0	77.7	106.0
45	23.5	0.637	56.4	77.7	103.6
50	21.2	0.655	56.3	78.0	101.7

(5) $p_y = 355 \, \text{N/mm}^2$ $f_c = 15 \, \text{N/mm}^2$

x	L_m/r_y	β_m	L'_m/r_y		
			$\beta = 0.25$	$\beta = 0$	$\beta = -0.75$
20	47.9	0.484	68.0	89.4	134.1
25	39.6	0.503	59.5	79.2	115.9
30	33.7	0.521	53.6	71.9	102.7
35	29.2	0.540	51.9	71.5	99.7
40	25.8	0.558	50.9	71.4	97.3
45	23.0	0.577	50.5	71.5	95.3
50	20.8	0.595	50.4	71.8	93.7

(6) $p_y = 355 \, \text{N/mm}^2$ $f_c = 30 \, \text{N/mm}^2$

x	L_m/r_y	β_m	L'_m/r_y		
			$\beta = 0.25$	$\beta = 0$	$\beta = -0.75$
20	44.0	0.424	58.5	79.2	118.9
25	37.4	0.443	52.2	71.5	104.6
30	32.3	0.461	47.5	65.5	93.5
35	28.3	0.480	46.0	65.4	91.2
40	25.1	0.498	45.2	65.4	89.2
45	22.6	0.517	44.8	65.6	87.5
50	20.5	0.535	44.7	65.9	86.0

μ = the coefficient of friction between concrete and steel
γ_f = the partial safety factor for load
D = the depth of the steel beam.

From BS 5950: Part 1, clause 4.3.2.1, $r = 2.5\% = 0.025$. In a typical beam with the live load approximately equal to the dead load and with a span to depth ratio $L/D = 25$, this gives

$$\mu > \frac{(1.4 + 1.6)}{2} \frac{25}{320} = 0.12$$

It is stated (Reference 3.17) that for an unpainted steel flange, μ is in the order of 0.3 while for a painted flange it usually exceeds 0.10. The requirements for member stability are therefore usually satisfied by this form of construction.

3.3.4.2   Stability of semi-continuous beams

Figure 3.25 shows a semi-continuous beam with both sagging and hogging bending moment zones. Usually the hogging plastic hinge M_1 at the support is the first hinge to form and requires a torsional

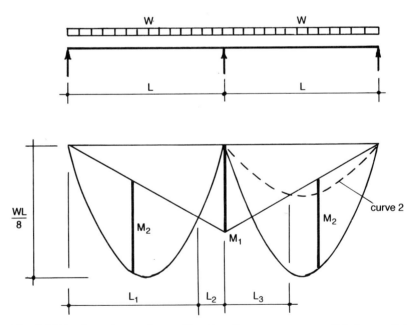

Fig. 3.25 Semi-continuous beam. Free bending moment curve 2 is for partial (pattern) loading; L_1 is the sagging moment zone; L_2 is the hogging moment zone for symmetrical loading; L_3 is the hogging moment zone for pattern loading.

restraint. This can be provided by the beam web stiffeners and the beam connection to the column cap plate. A moment connection to the column flange would not normally be recommended because this connection would have to be designed to provide a plastic hinge with rotation capacity, or else sufficient strength to ensure that the hinge formed solely in the beam.

The sagging hinge M_2 is normally the second plastic hinge to form and does not need a torsional restraint. The second hinge may well not have formed at the required ultimate limit state load because the choice of discrete steel sections usually ensures some spare capacity, albeit small. To take advantage of this the designer has to show that the sagging hinge is the last hinge to form. This is a simple operation using an elastic analysis.

As discussed in Section 3.3.3 above, the checking of the stability of the hogging moment length L_2 is not adequately covered in BS 5950 because in this case the bending moment diagram profile is advantageously not constant. Another point to bear in mind is that hogging moment length increases considerably under pattern loading. A realistic pattern loading should therefore be assumed.

The positions of the points of contraflexure shown in Fig. 3.25 are as follows:

Full loading on both spans $\hphantom{xxxxxxxxxxxxxxxxxxxxxxxxxx}$ $L_2 = 0.1715L$
Pattern loading $\hphantom{xxxxxxxx}$ Dead/imposed $= 1.0$ $\hphantom{xx}$ $L_3 = 0.3675L$
$\hphantom{xx}$ (imposed on only one span) $\hphantom{xx}$ Dead/imposed $= 0.5$ $\hphantom{xx}$ $L_3 = 0.5635L$

For each of these point of contraflexure positions, the shape of the bending moment diagram is basically the same. There is a plastic hinge at one end and zero bending moment at the other end with a slightly curved distribution in between. Assuming a linear bending moment in between is only a little conservative and the method given in Section 3.3.3 can be used with β, the ratio of smallest end moment to the largest end moment, equal to 0 (see Fig. 3.22).

A lateral restraint is required at a position adjacent to the point of contraflexure. With a beam supporting precast concrete units, this can be provided by providing small vertical steel plates welded to the top flange of the beam together with utilisation of the local continuity reinforcement tying the units together. These vertical plates spaced regularly along the beam provide a good Health and Safety feature in that they help in the positioning of the precast concrete units and reduce the risk of the units slipping off one end bearing. Special steel restraint members would require fire protection to match that of the steel floor beams.

Table 3.3 Maximum allowable distance between restraints with $\beta = 0$.

Torsional index (Universal Beam section)	$p_y = 275\,\text{N/mm}^2$		$p_y = 355\,\text{N/mm}^2$	
	L_m from clause 5.3.5	L'_m from Section 3.3.3	L_m from clause 5.3.5	L'_m from Section 3.3.3
20	$68r_y$	$132r_y$	$53r_y$	$102r_y$
25	$55r_y$	$114r_y$	$42r_y$	$88r_y$
30	$46r_y$	$102r_y$	$35r_y$	$79r_y$
40	$34r_y$	$100r_y$	$26r_y$	$78r_y$
50	$27r_y$	$101r_y$	$21r_y$	$78r_y$

The verification of the stability of lengths L_2 and L_3 is best done starting with the simple and fast-to-use formulae and progressing to the more complicated ones if required as listed below.

(1) Clause 5.3.5 of BS 5950: Part 1   This disadvantageously takes $\beta = 1.0$
(2) Section 3.3.3 in this book   with $\beta = 0$
(3) Reference 3.16   for non-linear bending moments
(refer to Table 3.2)

A comparison between (1) and (2) is given (for zero axial stress f_c) in Table 3.3.

3.3.4.3 Example 3.1: Beam with discrete load points and restraint positions

Figure 3.26 shows a semi-continuous beam with point loads at the third points. Restraints can be provided at the load positions.

The three lengths AB, BC and CD have very different bending moment profiles. Assuming that the beam is of steel Grade 43 and has a torsional index x of 30, then the following conclusions can be drawn. Length AB has a bending moment profile similar to that considered in the previous example with the end moment ratio β equal to 0. Hence, from Table 3.2, $L'_m \le 102r_y$. Length BC has virtually a fairly constant bending moment profile with an end moment ratio $\beta = 0.666$. Hence a study of Section 3.3.3 will quickly show that L_m cannot be increased above the limit in clause 5.3.5 so that $L_m \le 45.6r_y$. Length CD has a better bending moment profile than AB and has an end moment ratio β of -0.666. The use of Section 3.3.3 with $f_c = 0$ gives the following:

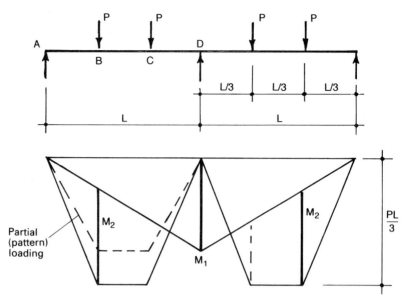

Fig. 3.26 Semi-continuous beam with discrete point loads.

$$K = 2.3 + 0.03 \times 30 = 3.2$$

$$K_0 = \frac{180 + 30}{300} = 0.7$$

$$L \le K\left[K_0 - \frac{\beta(1 - K_0)}{0.75}\right] L_m = 3.2\left[0.7 + \frac{0.666(1 - 0.7)}{0.75}\right] 45.6 r_y$$

$$= 141 r_y$$

This result could, of course, have been obtained more easily by linear interpolation from Table 3.2.

This example again emphasises the point that members like BC which have plastic hinges in zones of near constant bending moment have severe stability problems. It is easy to show that the plastic hinge at B is the last one to form and therefore it is practical to select a steel section that ensures that at the required design load the beam is still elastic at position B. Then member BC (and AB) could be checked for elastic stability to BS 5950: Part 1, clause 4.3.7.

3.3.5 *Portal frame member stability*

In general, the member stability checks should be carried out using the final plastic bending moment diagram for the loading combination used to determine the member sizes. Usually, the critical loading combination is $1.4 \times$ Dead $+ 1.6 \times$ Snow $+$ Notional horizontal loads (0.5% of the factored Dead $+$ Snow loads). It is normal practice to use this combination for member sizing and for positioning the restraints. The resulting design is then checked for adequacy under other loading combinations, if considered necessary. The notional horizontal loads usually have a less than 1% effect on the required plastic moments of resistance and can be ignored in member sizing.

The other loading combinations most likely to require consideration for member stability checks are:

$$1.0 \text{ Dead} + 1.4 \text{ (transverse wind} + \text{internal pressure)}$$

$$\text{and} \quad 1.0 \text{ Dead} + 1.4 \text{ (longitudinal wind} + \text{internal pressure)}$$

These two cases can sometimes put the rafter under net uplift for at least half of the portal span so that a long length of the unrestrained bottom flange of the rafter is under compression. However, in these two cases it is also very likely that the frame will be completely elastic at the required ultimate limit state loading.

When using rigid plastic design by non-computer methods, it is usual to produce a plastic bending moment diagram for the frame at collapse. This will reflect the excess capacity due to the choice of discrete member sizes in some arbitrary way. Ideally, the last plastic hinge to form should be identified. With the semi-graphical methods used in Chapters 5 and 7, the reactant bending moment diagram can then be adjusted so that the excess capacity is associated with the last hinge to form.

With elastic-plastic analysis, using a computer program, it is usual to produce an elastic-plastic bending moment diagram for the precisely required loading at the ultimate limit state. Hence only a partial collapse mechanism is likely to exist and there are fewer plastic hinges requiring torsional restraint. This also usually results in smaller bending moments in the centre span portion of the rafter, thus further alleviating the member stability situation.

It is important to realise that the elastic-plastic bending moment diagram at the design load is not an exactly scaled down version of the diagram that gives the full collapse mechanism in conjunction with the excess member capacity.

The eaves plastic hinge can form either in the column or in the rafter adjacent to the toe of the eaves haunch. Although it is not stated in BS 5950, it is not recommended to optimise the design in order to achieve both hinges occurring simultaneously because this adversely affects the stability of the haunch. If the designer chooses to optimise in this way and to use a haunch length that achieves these two hinges, then the stability of the eaves haunch requires very careful consideration.

It is recommended that designers use the procedures described in Chapters 5 and 7 whereby the eaves hinge is located in the column below the haunch and the length of the haunch is chosen so that the toe of the haunch is completely elastic. This minimises the sometimes troublesome problems of stabilising the haunch and rafter.

3.3.5.1   The point of contraflexure as a position of restraint

It is no longer satisfactory to automatically assume that a point of contraflexure is a virtual position of lateral restraint. Before this can be done, it is necessary that all of the purlin-rafter connections comply with the requirements of clause 5.5.3.5.2, namely:

❑ every length of purlin has at least two bolts in each purlin-rafter connection; and
❑ the depth of the purlin section is not less than 0.25 times the depth of the rafter.

This is to ensure a notional adequate torsional stiffness at every purlin connection and to reflect the situation in the structural tests from which the stability requirements were derived.

When the above criteria are not met then an effective lateral restraint to the bottom flange must be used if a position of restraint is to be assumed in the member stability calculations.

It should be remembered that the position of the point of contra-flexure varies continuously as the frame changes from elastic behaviour through the formation of successive plastic hinges into a plastic collapse mechanism. Care should be taken to ensure that the position chosen, and the corresponding bending moment diagram, represent the worst design conditions for the member being considered.

3.3.5.2   Restraints in portal frame construction

Purlins and sheeting rails are assumed to provide lateral restraint to the flange to which they are connected. On their own, they do not provide either torsional restraint to the section or lateral restraint to the remote flange if this flange is in compression.

The most usual way of achieving either of these forms of bracing is to use a knee brace (otherwise known as a fly brace) connected to a purlin or sheeting rail as shown in Fig. 3.27. Research[3.18] indicates that each knee brace should be designed for a compression load equal to 2% of the compression flange yield load and should have a stiffness that is given by a slenderness ratio of at least 100.

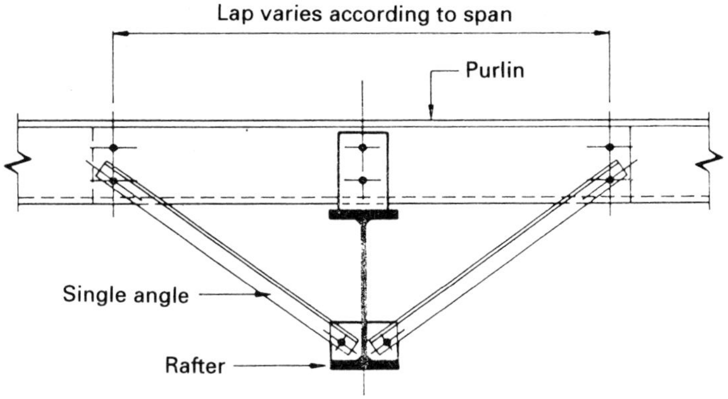

Fig. 3.27 Typical knee brace.

It should not be assumed that knee braces connected to cold-formed purlins or rails provide adequate restraint irrespective of the size of the restrained member. Large rafters or columns (in the order of 686 × 254 UB) require substantial direct bracing systems. On the other hand, small rafters and columns (in the order of 254 × 146 UB), as an alternative to knee braces, probably only need full depth web stiffeners adjacent to the purlin-rafter connections (which should comply with clause 5.5.3.5.2 as described above).

3.3.5.3 *Procedures for checking member stability in portal frames*
An overview of the relevant clauses and formulae has been given in Section 3.3.1. This section outlines how these are usually applied to the members of portal frames.

Torsional restraints are required at all plastic hinge locations except at the last hinge to form, provided that it can be clearly identified. Under uniform vertical loading of a single bay frame, hinges usually form in symmetrical pairs and this statement applies to the last pair to form. When an elastic-plastic analysis is available, it is not necessary to provide torsional restraint at any hinge which forms at a load above the factored design load. The advantages of identifying plastic hinges which are part of the collapse mechanism but which do not require torsional restraint cannot be overstressed.

There are then three different regions of the frame which require checking and it is convenient to consider these one at a time.

The apex region of the rafter contains a length of almost uniform bending moment and this is the most difficult distribution to stabilise. However, the most critical load condition places the outer flange in compression and this is laterally restrained by the purlins. The requirements therefore depend critically on whether it has been possible to prove that this part of the frame is entirely elastic at the ultimate limit state. If this part of the rafter contains a plastic hinge, then the purlin spacing is limited to L_m in clause 5.3.5. If this part of the rafter is elastic, and in any event for parts of the rafter remote from the plastic hinge, then the standard requirements for elastic design given in Section 4.8.3 of BS 5950 apply. The purlin spacing is then restricted to a value of L_E satisfying this part of the Standard.

In the haunch region of the rafter the purlins restrain the tension flange. If this region contains a plastic hinge, then the purlin spacing in the vicinity of the hinge is again limited by L_m but, according to clause 5.5.3.5.2, knee braces or other braces to the inside flange are only required at a spacing given conservatively by L_s in this clause. Provided that the conditions described in Section 3.3.5.1 above are satisfied, the bottom flange can also be assumed to be restrained at the point of contraflexure. If this does not result in a satisfactory design, then Appendix G can be tried.

If the advice given in this book has been followed, then the haunch region of the rafter will generally be elastic. In this case the maximum purlin spacing can be increased to that which results in a value of L_E which satisfies Section 4.8.3 of BS 5950. The requirements for the spacing of restraints to the inside compression flange remain as either L_s according to clause 5.5.3.5.2 or a value which satisfies the requirements of Appendix G.

Under the dominant load condition, *the column* will also usually be restrained at intervals along the tension flange by the sheeting rails. In a plastically designed frame, there will almost always be a plastic hinge below the haunch. This will, of course, require a torsional restraint. According to BS 5950, the distance to the next restraint should be L_m in clause 5.3.5. However, the bending moment profile is not constant and a better answer may be given by using L'_m according to Section 3.3.3 of this book.

Turning now to the *stanchion region*, strictly speaking, clause 5.5.3.5 should only be applied to portal frame *rafters* but there seems to be no rational reason why it should not also be applied to stanchions restrained by sheeting rails. If this is accepted, then clause 5.5.3.5 states that the required restraints can either be torsional restraints or they can be tension flange restraints satisfying clause 4.8.3.3.1. In any case, a torsional restraint is, of course, required within a distance L_s of the plastic hinge. As an alternative, if there are rails present, Appendix G provides a more detailed approach to the design of this part of the structure.

If the column is elastic at the ultimate limit state, and for the lengths which are remote from the plastic hinge, the procedure is again similar to that described above with the difference that the distance between restraints is not governed by L_m or L'_m but rather by L_E in section 4.8.3 of BS 5950.

3.3.6 The stability of members in multi-storey frames

The stability of beams in multi-storey construction can be checked by the methods described in Section 3.3.4 of this book. Columns should be checked in accordance with clause 4.8.3 if they are shown to be elastic by the frame design requirements in BS 5950. Refer to Chapter 6 in this publication.

Columns containing plastic hinges cannot be effectively checked because clause 5.3.5 is too conservative. Reference 3.15 is probably the best available method. It should be noted that the method described in Section 3.3.3 is also appropriate but *only* for Universal Beams. It cannot be used for multi-storey frames that have Universal Columns as stanchions.

References

3.1 Davies J M. Collapse and shakedown loads of plane frames, *Proc. ASCE, J Struct. Div.* ST3, June 1967, pp. 35–50.

3.2 Horne M R. The effect of variable repeated loads in building structures designed by the plastic theory. *Publ. Int. Ass. Bridge and Struct. Engng.*, Vol. 14, 1954, p. 53.

3.3 Davies J M. The response of plane frameworks to static and variable repeated loading in the elastic-plastic range. *Structural Engineer*, Vol. 44, No. 8, August 1966, pp. 277–83.

3.4 Davies J M. Variable repeated loading and the plastic design of structures. *Structural Engineer*, Vol. 48, No. 5, May 1970, pp. 181–94.

3.5 Davies J M. Approximate minimum weight design of steel frames. *Proc. Int. Symp. Computer-aided Structural Design*, University of Warwick, July 1972, Peter Peregrinus, pp. B1.16–B1.32.

3.6 Horne M R. Elastic-plastic failure loads of plane frames. *Proc. Roy. Soc. A*, Vol. 274, 1963, p. 343.

3.7 Low M W. Some model tests on multi-storey rigid steel frames. *Proc. ICE*, Vol. 13, 1959, p. 287.

3.8 Horne M R. An approximate method for calculating the elastic-critical loads of multi-storey plane frames. *Structural Engineer*, Vol. 53, No. 6, June 1975.

3.9 Horne M R. Safeguards against frame instability in the plastic design of single-storey pitched roof frames. *Proc. Conf. on the Behaviour of Slender Structures*. The City University, London, 1977.

3.10   Davies J M. In-plane stability in portal frames. *Structural Engineer*, Vol. 68, No. 8, April 1990, pp. 141–7.

3.11   Davies J M. The stability of multi-bay portal frames. *Structural Engineer*, Vol. 69, No. 12, June 1991, pp. 223–9.

3.12   Coates R G, Coutie M G & Kong F K. Structural analysis, 3rd ed., Van Nostrand Reinhold (UK), 1988.

3.13   Advisory Desk AD 097. Nominal base stiffness. *Steel Construction Today*, November 1991, pp. 285–6.

3.14   Plastic design of single-storey pitched-roof portal frames to Eurocode 3. Steel Construction Institute Technical Report 147, 1995.

3.15   Horne M R. Safe loads on I-section columns in structures designed by plastic theory. *Proc. ICE*, Vol. 29, 1964, pp. 137–50.

3.16   Brown B A. The requirements for restraints in plastic design to BS 5950. *Steel Construction Today*, Vol. 2, No. 6, December 1988, pp. 184–6.

3.17   Nethercot D A & Lawson R M. Lateral stability of steel beams and columns – common cases of restraint. Steel Construction Institute Publication No. 093, 1992.

3.18   Morris L J & Plum D R. Structural steelwork design to BS 5950. Longmans Scientific and Technical, 1988.

Chapter 4
Plastic Design of Beams

4.1   General

Many floor and roof beams support uniformly distributed dead and imposed loads. The design depends crucially on whether the beam is restrained or unrestrained. A restrained beam will have its compression flange restrained against lateral movement so that the section can develop its full plastic bending moment capacity. An unrestrained beam may be subject to lateral torsional buckling and therefore may not be able to develop its full plastic moment of resistance.

Clause 4.2.2 of BS 5950: Part 1: 1990 defines the requirements for full lateral restraint such that lateral torsional buckling need not be considered as follows:

'Full restraint exists if the frictional or positive connection of a floor or other construction to the compression flange of the member is capable of resisting a lateral force of not less than 2.5% of the maximum factored force in the compression flange of the member under factored loading.

This load should be considered as distributed uniformly along the flange, provided that the dead load of the floor and the imposed load it supports together constitute the dominant loading on the beam. The floor construction should be capable of resisting this lateral force.'

The requirements for a frictional connection between a beam and the floor that it supports have been considered in Section 3.3.4.1 of this book. It follows that, in general, a beam that continuously supports a floor or roof will be restrained. Conversely, all other types of bending member are likely to require checking for lateral stability.

The plastic design of beams is straightforward using the procedures described in Chapter 2 together with consideration of member stability according to Chapter 3. This chapter will therefore concentrate on two examples which will draw together the above points and introduce some secondary considerations.

From the point of view of practical design, continuous beams can be advantageously designed by combining the free bending moment diagrams of each individual span with a suitable reactant moment diagram to give a collapse mechanism in the most critical span. The collapse mechanism is partly under the control of the designer because variations in beam section sizes and splice positions will help to determine the critical span and the collapse mechanism.

4.2   Example 4.1. Continuous beam of uniform section

The continuous beam shown in Fig. 4.1 will be designed as one continuous section. It is unrestrained except at the load points.
  The point loads P at the ultimate limit state arise as follows:

Dead load $= 100$ (unfactored) $\times 1.4$ $= 140\,\text{kN}$
Imposed load $= 150$ (unfactored) $\times 1.6$ $= 240\,\text{kN}$

Total design load $= 380\,\text{kN}$

Hence the free bending moment diagram shown in Fig. 4.2(a) can be drawn.

$$\text{For span AB,} \quad \frac{PL}{4} = \frac{380 \times 6}{4} = 570\,\text{kNm}$$

$$\text{For span BC,} \quad \frac{PL}{2} = \frac{380 \times 6}{2} = 1140\,\text{kNm}$$

4.2.1   Choice of section size

Once the free bending moment diagram has been drawn, the use of one continuous beam section means that the critical span can usually be

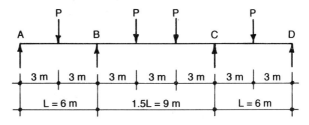

Fig. 4.1 General arrangement of the continuous beam.

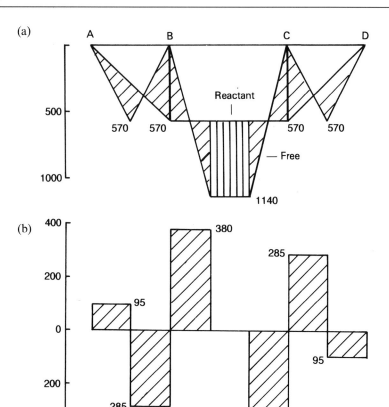

Fig. 4.2 Bending moment and shear force diagrams for continuous beam: (a) bending moment diagram (kNm); (b) shear force diagram (kN).

found by inspection bearing in mind that the end spans are simply supported at one end and the reactant moment must reduce to zero at the two ends of the structure. Here, it is clear that the centre span BC is critical with plastic hinges at B and C and a long region of full plasticity between the loads.

The reactant bending moment line can now be added to Fig. 4.2(a) for a member with a plastic moment of resistance given by $M_p = 1140/2 = 570$ kNm. Assuming a steel design grade 43, then $p_y = 275$ N/mm^2 (or less for thicker sections) and the required section plastic modulus $S_x = 570/0.275 = 2073$ cm^3. Hence a suitable Universal Beam section is a $533 \times 210 \times 92$ UB with $S_x = 2366$ cm^3. This is a Class 1 section in accordance with clause 3.5.2.

4.2.2   Effect of shear forces

From the combined free and reactant bending moment diagrams in Fig. 4.2(a), the net bending moments are available and the corresponding reactions at A,B,C and D are statically determinate. The shear force diagram can therefore be derived as shown in Fig. 4.2(b).

Clauses 4.2.5 and 4.2.6 of BS 5950: Part 1 define shear loads as either low shear loads or high shear loads. It is only the high shear loads that reduce the plastic moment of resistance of the beam.

Clause 4.2.3 defines the shear load capacity P_v of the section where

$$P_v = 0.6 p_y D t$$

and where D is the overall depth of the section
t is the thickness of the web

Values for P_v are tabulated in *Steelwork Design*[4.1] where, on page 139, for a 533 × 210 × 92 UB, $P_v = 897$ kN.

An applied shear load F_v is classified as a low shear load if $F_v \le 0.6 P_v$ and here,

$$0.6 P_v = 0.6 \times 897 = 538 \text{ kN}$$

It can be seen that this is not exceeded anywhere on the shear force diagram in Fig. 4.2(b) so that no reduction in the moment capacity of the beam section is necessary.

4.2.3   Plastic hinge history

It is neither usual nor necessary to determine the plastic hinge history but, if this can be done, it does give the advantage of checking the stability of the beam members at the minimum required ultimate limit state loading rather than at the usually higher collapse mechanism loading.

In this example, member BC has a plastic hinge along the entire length between the loads and, as mentioned in Section 3.3 of Chapter 3, this creates a severe instability problem. However, if this hinge is the last hinge to form, or if it can be shown that it has not formed at the minimum required ultimate limit state loading, then the stability criteria are less severe.

A simple elastic analysis carried out for the arbitrary load $P = 100$ kN gives the bending moment diagram shown in Fig. 4.3. The largest bending moment is at supports B and C and is 173.1 kNm.

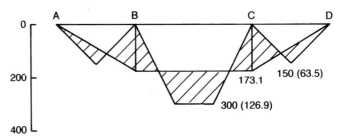

Fig. 4.3 Elastic bending moment diagram (kNm) for $P = 100\,\text{kN}$.

Therefore the first plastic hinges form simultaneously at B and C. The plastic moment of resistance of the chosen section is

$$M_p = p_y S_x = 0.275 \times 2366 = 650.6\,\text{kNm}$$

a result which could also have been obtained from page 129 of Reference 4.1. It follows that the load at which the first hinge forms is

$$P = \frac{100 \times 650.6}{173.1} = 375.9\,\text{kN}$$

If the loading is increased further, positions B and C behave as pins while retaining the fully plastic moment of the section. Beams AB, BC, and CD then act as though they were simply supported as far as any additional load is concerned and additional sagging moments are added to the potentially critical hinge positions under the loads. It can be seen that the total moment at mid-span of BC will reach the full plastic moment of the section before any additional plastic hinge forms in the outer spans AB and CD. Hence the next hinge results in a beam mechanism in span BC, as shown in Fig. 4.4. The load at which the second hinge forms is

$$P = \frac{650.6 \times 2}{3} = 433.7\,\text{KN}$$

The mid-span deflection of beam BC at the yield load $P = 375.9\,\text{kN}$ given by the elastic analysis is 27.7 mm. The deflection δ_c at collapse is

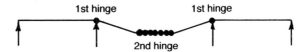

Fig. 4.4 Collapse mechanism for continuous beam.

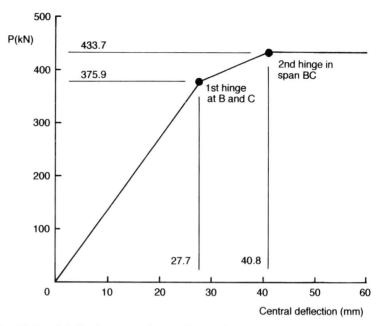

Fig. 4.5 Load-deflection curve for continuous beam.

found by adding to this value the deflection of the middle span acting as a simply supported beam under the additional load of $433.7 - 375.9 = 57.8\,\text{kN}$, thus

$$\delta_c = 27.7 + \frac{23 \times 57.8 \times (1.5L)^3}{648\,EI}$$

$$= 27.7 + \frac{23 \times 57.8 \times 9000^3}{648 \times 205 \times 5.533 \times 10^8} = 40.8\,\text{mm}$$

Hence the load deflection curve and plastic hinge history are as shown in Fig. 4.5. The bending moment at the design load of $P = 380\,\text{kN}$, which will be used for the member stability checks, is shown in Fig. 4.6.

4.2.4 Member stability

4.2.4.1 Left-hand half of member AB
This length of member remains elastic throughout and can be checked in accordance with clause 4.3.7. Under the load,

$$M_{AB} = \frac{380 \times 6}{4} - \frac{650.6}{2} = 244.7\,\text{kNm}$$

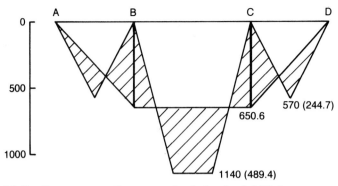

Fig. 4.6 Bending moment diagram at the design load (kNm).

From Table 18, for an end moment ratio $\beta = 0$, $m = 0.57$, so that

$$\bar{M} = mM_{AB} = 0.57 \times 244.7 = 139.5 \, \text{kNm}$$

This must be less than $M_b = S_x p_b$ where p_b is from Table 11 and is based on λ_{LT} where

$$\lambda_{LT} = nuv\lambda$$

From Table 13, $n = 1.0$ and from *Steelwork Design*[4.1] for a $533 \times 210 \times 92 \, \text{UB}$, $u = 0.871$.

$$\lambda = \text{minor axis slenderness ratio} = \frac{0.85 \times 3000}{45.0} = 56.7$$

From Reference 4.1, $x = 36.4$ and hence $\lambda/x = 1.56$ giving, from Table 14, $v = 0.97$ and hence

$$\lambda_{LT} = 1.0 \times 0.871 \times 0.97 \times 56.7 = 47.9$$

The required value of the bending strength p_b now follows by entering Table 11 for $p_y = 275 \, \text{N/mm}^2$, giving $p_b = 243 \, \text{N/mm}^2$.

It follows that the buckling resistance moment $M_b = 2366 \times 0.243 = 575.0 \, \text{kNm}$ which is greater than the equivalent uniform bending moment of 139.5 kNm and the member is stable. This same value of $M_b = 575 \, \text{kNm}$ can also be obtained directly by linear interpolation of the table on page 129 of Reference 4.1.

4.2.4.2   Right-hand half of member AB
This member has a plastic hinge (the first to form) at end B and will therefore be checked in accordance with Section 3.3.3 of this book.

$$\beta = \frac{-244.7}{650.6} = -0.376 \quad \text{and the axial stress } f_c = 0.0$$

$$L_m = \frac{38r_y}{\left[\dfrac{f_c}{130} + \left(\dfrac{p_y}{275}\right)^2\left(\dfrac{x}{36}\right)^2\right]^{0.5}} = \frac{38 \times 45.0}{\left(\dfrac{36.4}{36}\right)} = 1691 \text{ mm}$$

$$\beta_m = 0.44 + \frac{x}{270} - \frac{f_c}{200} = 0.44 + \frac{36.4}{270} = 0.575$$

$$K = 0.8 + 0.08x - \frac{(x-10)f_c}{2000} = 0.8 + 0.08 \times 36.4 = 3.712$$

$$K_0 = \frac{180 + x}{300} = \frac{180 + 36.4}{300} = 0.721$$

$$L'_m = K\left[K_0 - \frac{\beta(1 - K_0)}{0.75}\right]L_m$$

$$= 3.712\left[0.721 - \frac{(-0.376)(1.0 - 0.721)}{0.75}\right]1691 = 5403 \text{ mm}$$

As this is greater than the length of 3000 mm between restraints, the member is stable.

4.2.4.3   Left-hand portion of member BC
This member is similar to the right-hand half of member AB and has a better β ratio.

$$\beta = \frac{-(1140 - 650.6)}{650.6} = -0.752$$

Therefore from Section 3.3.3, $L = KL_m = 3.712 \times 1691 = 6276$ mm which is greater than 3000 mm and the member is stable.

4.2.4.4   Central portion of member BC
In the presence of a complete collapse mechanism, this part of the member would have a plastic hinge along its complete length. From the checks carried out on other members, it is obvious that the stability of this member at collapse could not be justified and therefore there is an advantage to be gained by checking it at the minimum required design loading for the ultimate limit state. This takes advantage of the excess capacity due to the choice of a discrete size of Universal Beam and the fact that the first plastic hinges to form are elsewhere at supports B and C.

At the minimum required design loading, this part of the member has a constant bending moment along its complete length of $1140 - 650.6 = 489.4\,\text{kNm}$ and this can be checked in accordance with clause 4.3.7 of BS 5950. From Table 18, for an end moment ratio $\beta = 1.0$, $m = 1.0$ and therefore $mM = 489.4\,\text{kNm}$.

From the calculations for member AB, $M_b = S_x p_b = 575\,\text{kNm}$ which is greater than $489.4\,\text{kNm}$ and therefore the central part of member BC is stable.

4.2.5   Effect of settlement at a support

Settlement at one or more supports affects the plastic hinge history but not the collapse load or the final collapse mechanism. Consider, as an illustration of this, the effect of settlement at support B on the hinge history of the continuous beam of Example 4.1. In the elastic range of loading, the deformation pattern shown in Fig. 4.7(a) and the corresponding bending moments shown in Fig. 4.7(b) are superimposed on the results of a conventional analysis without settlement.

For the state of affairs shown in Fig. 4.7, it can be shown that

$$M_1 = 0.30RL$$

$$M_2 = 0.48RL$$

$$\delta = \frac{0.195RL^3}{EI}$$

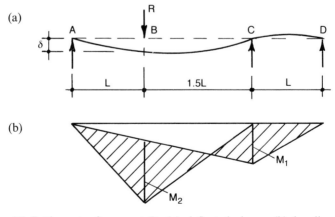

Fig. 4.7 Settlement of support B: (a) deflected shape; (b) bending moment diagram.

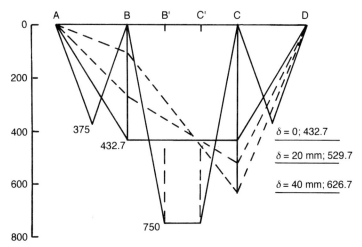

Fig. 4.8 Effect of settlement at B on the elastic bending moments (kNm).

For the beam under consideration, $I = 55\,330\,\text{cm}^4$, $L = 6000\,\text{mm}$ and $E = 205\,000\,\text{N/mm}^2$ and it follows that for a settlement $\delta = 10\,\text{mm}$, $R = 26.9\,\text{kN}$, $M_1 = 48.5\,\text{kNm}$ and $M_2 = 77.6\,\text{kNm}$.

Hence, in Fig. 4.8, the free and reactant bending moments are shown at the serviceability load of $P = 250\,\text{kN}$ for the beam without settlement and for 20 mm increments of δ. As the settlement at B increases, the bending moment at C increases. In order for a plastic hinge to occur at C at the serviceability limit (unfactored load) of $P = 250\,\text{kN}$ the settlement δ is given by

$$\delta = \left(\frac{650.6 - 432.7}{48.47} \right) 10 = 45.0\,\text{mm}$$

For higher values of settlement than this, the first plastic hinge can form at loads below the serviceability load of the structure.

Pursuing the hinge history for this value of δ by a series of elastic analyses as summarised in Fig. 4.9, the behaviour up to collapse can be traced. Further details of this analysis are given in Table 4.1.

Comparing Table 4.1 with Fig. 4.5, it can be seen that the settlement causes a completely different plastic hinge history but the final collapse load and collapse mechanism are unchanged.

4.3   Example 4.2. Non-uniform section beam

The geometry and loading for this example will be exactly the same as were used for Example 4.1. However, the critical central region of span BC will be strengthened by flange plates resulting in a more economical

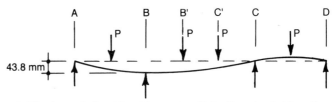

(a) Case 1. Elastic analysis with settlement until the first plastic hinge forms at C

(b) Case 2. Elastic analysis with a pin at C until second hinge forms at B′

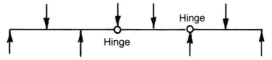

(c) Case 3. Elastic analysis with pins at B′ and C until final hinges form at B and C′

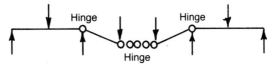

(d) Case 4. Plastic collapse mechanism as before

Fig. 4.9 Steps in following the hinge history after settlement at support B.

section for the remainder of the beam. The beam and its design loading for the ultimate limit state are shown in Fig. 4.10.

The free bending moment diagram is unchanged from Example 4.1 and the design is started by adding the reactant moment diagram that achieves collapse mechanisms in the end spans as shown in Fig. 4.11. It can then be seen where the beam has to be strengthened in order to carry the bending moments which exceed the capacity of the section.

From Fig. 4.11, a steel beam with a full plastic moment of 380 kNm is required and, if steel design Grade 43 is used, the required plastic

Table 4.1 Plastic hinge history of continuous beam with settlement of B.

Step	ΔP	P	M_B	M'_B	M'_C	M_C
1	250.0	250.0	−84.0	477.1	288.2	−650.6
2	115.7		−260.3	173.5	260.3	0
		365.7	−344.3	650.6	548.5	−650.6
3	68.0		−306.3	0	102.1	0
		433.7	−650.6	650.6	650.6	−650.6

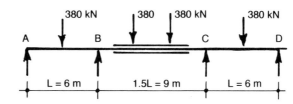

Fig. 4.10 Non-uniform continuous beam.

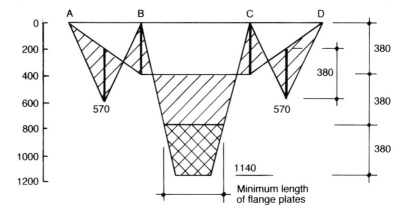

Fig. 4.11 Bending moment diagram (kNm) for collapse of the end spans.

modulus is $380/0.275 = 1382 \, \text{cm}^3$. Hence a suitable Universal Beam section is a $457 \times 191 \times 67$ with $S_x = 1472 \, \text{cm}^3$. This is a Class 1 section in accordance with clause 3.5.2. The plastic moment of resistance $= 0.275 \times 1472 = 404.8 \, \text{kNm}$.

4.3.1 Effect of shear forces

The shear force diagram is shown in Fig. 4.12.

From page 139 of *Steelwork Design*[4.1], the shear load capacity of the chosen section is $P_v = 636 \, \text{kN}$ and for low shear load classification according to clause 4.2.5 the applied load F_v should be less than $0.6 P_v = 0.6 \times 636 = 381.6 \, \text{kN}$. This load is not exceeded anywhere in the shear force diagram and therefore there is no reduction in the beam moment capacity due to shear loads.

4.3.2 Strengthened beam section

From Fig. 4.11, the strengthened beam section requires a full plastic moment of 760 kNm and therefore the required plastic section modulus is $760/0.275 = 2764 \, \text{cm}^3$.

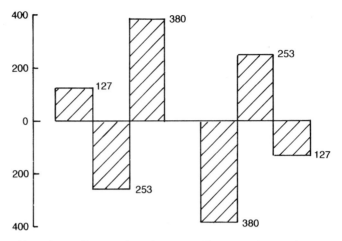

Fig. 4.12 Shear force diagram (kN) for non-uniform continuous beam.

Flange plates are required to achieve $S_x = 2764 \, \text{cm}^3$. A minimum width plate has to accommodate rolling and straightness tolerances and welds. It should be readily available and economic and should therefore be chosen from the standard range. Hence a width of 250 mm is selected.

The minimum plate thickness has to achieve Class 1 to Table 7 of BS 5950 which means that $b/T \le 7.5$ for outstand elements and ≤ 23 for internal elements respectively. Hence a thickness of 12 mm is selected. The chosen flange plate arrangement is shown in Fig. 4.13. The b/T ratio of the internal element is $189.9/12 = 15.8$. The outstand element is clearly not critical.

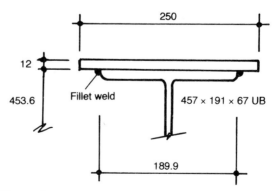

Fig. 4.13 Flange plate.

The section properties of the strengthened section are as follows:

$$S_x = 1472 + 2 \times 25 \times 1.2 \times 23.28 = 1472 + 1396 = 2868\,\text{cm}^3$$

$$I_x = 29\,410 + 2 \times 25 \times 1.2 \times 23.28^2 = 29\,410 + 32\,517 = 61\,927\,\text{cm}^4$$

The available section modulus of 2868 cm³ is greater than the required value of 2764 cm³ and therefore strength of the plated section is adequate.

4.3.3   Flange plate curtailment position

Figure 4.14 shows part of the preliminary bending moment diagram at collapse from which the flange plate curtailment position is determined.

A plastic hinge is not desirable in the basic beam section at the curtailment position because it would require restraints that are not readily available. Hence the bending moment at that position will be limited to the yield moment of the section at the required design load of $P = 380\,\text{kN}$.

$$M_e = p_y Z_x = 0.275 \times 1297 = 356\,\text{kNm}$$

$$\text{hence} \quad C = \frac{(1140 - 380 - 356)}{1140} \times 3.0 = 1.063\,\text{m}$$

The flange plates have to extend past the theoretical curtailment position given by C and this extension length can be determined from the weld strength calculations.

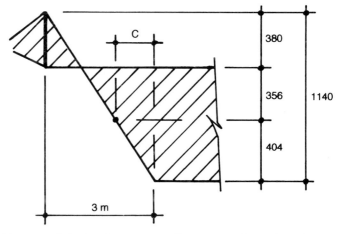

Fig. 4.14 Part of the bending moment diagram at collapse (kNm).

From Fig. 4.12, the relevant shear load in the beam is $Q = 380\,\text{kN}$

$$\therefore \quad \text{the weld shear load} = \frac{QAy}{I}$$

$$= \frac{380 \times 250 \times 12 \times 227}{61\,927 \times 10^4}$$

$$= 0.42\,\text{kN/mm} = 0.21\,\text{kN/mm per weld}$$

A 6 mm leg fillet weld has a capacity of 0.903 kN/mm (page 245 of Reference 4.1) and is more than adequate.

The load in the flange plate at the curtailment position is the load that the plate attracts from the bending moment of $M_e = 356\,\text{kNm}$.

$$\text{Thus the flange load} = \frac{M}{Z} \times \text{plate area}$$

$$= \frac{356 \times 239 \times 250 \times 12}{61\,927 \times 10} = 412\,\text{kN}$$

Based on two 6 mm fillet welds the required extension length

$$\frac{412}{2 \times 0.903} = 228\,\text{mm}$$

Hence the total length that the flange plates must extend past the point loads in span BC is $1063 + 228 = 1291\,\text{mm}$. They can therefore conveniently be continued to the middle of the 3 m length.

4.3.4   Plastic hinge history

As in Example 4.1, the plastic hinge history is useful for reasons of member stability and therefore economy. Figure 4.15 shows the beam which must be analysed in order to determine the first plastic hinge to form. It is more complicated than the beam in Example 4.1 because it is necessary to take account of the non-uniform sections.

For the basic beam $I_x = 29\,410\,\text{cm}^4$ and for the plated beam $I_x = 61\,927\,\text{cm}^4$. The value of the support moment M is relatively insensitive to the value of C, thus:

C	=		M	=	
		0.333L			0.243 P L
		0.4L			0.242 P L
		0.5L			0.241 P L

If the plates are conveniently curtailed midway between the loads and the supports, as determined above, $C = 0.5L$ and $M = 0.241PL = 0.241 \times 250 \times 6 = 361.5\,\text{kNm}$ under the unfactored loads.

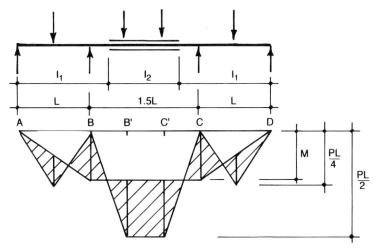

Fig. 4.15 Elastic analysis of plated beam.

The first plastic hinges form simultaneously at supports B and C at a load factor of

$$\frac{\text{Basic beam } M_p}{M} = \frac{404.8}{361.5} = 1.12$$

There are therefore no plastic hinges at the serviceability limit state.

For subsequent loading, the plastic hinges at B and C behave like pins and the three spans are independent of each other. At the minimum ultimate limit state loading of $P = 380$ kN, no further plastic hinges have formed and the bending moment diagram is as shown in Fig. 4.16. It is this bending moment diagram which will be used for the member stability checks.

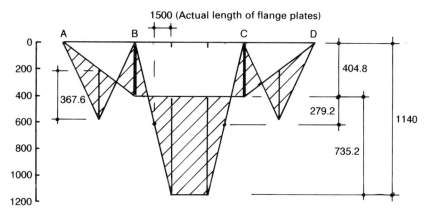

Fig. 4.16 Bending moment diagram (kNm) at ultimate limit state.

4.3.5 Member stability

4.3.5.1 Left-hand half of member AB

This length of member remains elastic throughout and can be checked in accordance with clause 4.3.7. Under the load,

$$M_{AB} = \frac{380 \times 6}{4} - \frac{404.8}{2} = 367.6\,\text{kNm}$$

From Table 18, for an end moment ratio $\beta = 0$, $m = 0.57$, so that

$$\bar{M} = mM_{AB} = 0.57 \times 367.6 = 209.6\,\text{kNm}$$

This must be less than $M_b = S_x p_b$ where p_b is from Table 11 and is based on λ_{LT} where

$$\lambda_{LT} = nuv\lambda$$

From Table 13, $n = 1.0$ and from *Steelwork Design*[4.1] for a $457 \times 191 \times 67$ UB, $u = 0.872$.

$$\lambda = \text{minor axis slenderness ratio} = \frac{0.85 \times 3000}{41.2} = 61.9$$

From Reference 4.1, $x = 37.9$ and hence $\lambda/x = 1.633$ giving, from Table 14, $v = 0.97$ and hence

$$\lambda_{LT} = 1.0 \times 0.872 \times 0.97 \times 61.9 = 52.4$$

The required value of the bending strength p_b now follows by entering Table 11 for $p_y = 275\,\text{N/mm}^2$ giving $p_b = 232\,\text{N/mm}^2$.

It follows that the buckling resistance moment $M_b = 1472 \times 0.232 = 341.5\,\text{kNm}$ which is greater than the equivalent uniform bending moment of 209.6 kNm and the member is stable.

This same value of $M_b = 341.5\,\text{kNm}$ can also be obtained directly by linear interpolation of the table on page 130 of Reference 4.1.

4.3.5.2 Right-hand half of member AB

This member has a plastic hinge (the first to form) at end B and will therefore be checked in accordance with Section 3.3.3 of this book.

$$\beta = \frac{-367.6}{404.8} = -0.908 \quad \text{and the axial stress } f_c = 0.0$$

$$L_m = \frac{38r_y}{\left[\frac{f_c}{130} + \left(\frac{p_y}{275}\right)^2 \left(\frac{x}{36}\right)^2\right]^{\frac{1}{2}}} = \frac{38 \times 41.2}{\left(\frac{37.9}{36}\right)} = 1487\,\text{mm}$$

$$\beta_m = \text{not relevant}$$

$$K = 0.8 + 0.08x - \frac{(x-10)f_c}{2000} = 0.8 + 0.08 \times 37.9 = 3.832$$

K_0 = not relevant

$$L'_m = KL_m = 3.832 \times 1487 = 5698 \text{ mm}$$

As this is greater than the length of 3000 mm between restraints, the member is stable.

4.3.5.3   Left-hand portion of member BC

This is a non-uniform member with a plastic hinge (the first to form) at end B.

Clause 5.3.5 of BS 5950: Part 1 is conservative for uniform members and is even more conservative for non-uniform members because it specifies that the worst section properties should be used in the equation for L_m.

A comparison with the calculations for the right-hand half of member AB is sufficient to indicate the nature of the problem that arises here. For a uniform member AB, L_m is only 1487 mm whereas 3000 mm is required. However, for AB with β at the most advantageous value of -0.908, $L'_m = 5698$ mm.

It is apparent that a decision based on engineering judgement has to be made in order to justify the stability of the left-hand part of BC and this would have to be agreed with the checking engineer. Various calculations might sufficiently indicate that the beam is stable. The options could be, for example:

(1)   The β ratio could be based on stresses, not bending moments, as shown in Fig. 4.17. In this case, the non-uniform member with a linear bending moment produces a non-linear stress distribution. Based on stresses, the β ratio is $-256/275 = -0.93$.

The r_y of the plated section is 5.61 cm compared with 4.12 cm for the basic Universal Beam. On the other hand the x value of the plated section is 38.1 (based on Appendix B.2.5 of BS 5950) compared with 37.9 for the basic Universal Beam. However, accounting for the non-linear stress distribution along the length of the member is not covered in BS 5950 and, although it appears to give a sensible answer, the use of the above values has no rigorous justification.

(2)   Formulae from other codes or sources may indicate the beam member is stable.

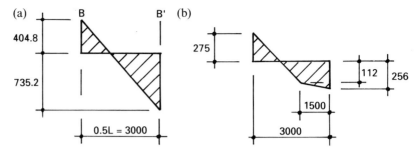

Fig. 4.17 Bending moments (kNm) and stresses (N/mm²) in member BC: (a) bending moment diagram (kNm); (b) stress diagram (N/mm²).

(3)   Intermediate restraints could be introduced.
(4)   The flange plates could be carried through to the support positions B and C.

Solutions (3) or (4) are, of course, more readily justifiable.

4.3.5.4   Central portion of member BC

At the minimum required ultimate limit state loading of $P = 380\,kN$ this length of member has a constant bending moment along its length of 735.2 kNm (from Fig. 4.16). The plastic moment of resistance of the plated beam $= S_x p_y = 2868 \times 0.275 = 788.7\,kNm$ and, as the member does not contain a plastic hinge requiring rotational capacity, it can be checked in accordance with clause 4.3.7 of BS 5950.

For a member under uniform bending moment, $\beta = 1.0$ and therefore $m = 1.0$ and $n = 1.0$. From BS 5950, Appendix B.2.5, $u = 0.914$ and $x = 38.1$

$$\therefore \quad \lambda = \frac{0.85 \times 3000}{56.1} = 45.5 \quad \text{and} \quad \frac{\lambda}{x} = 1.20$$

It follows from Table 14 that $v = 0.98$

$$\therefore \quad \lambda_{LT} = nuv\lambda = 1.0 \times 0.914 \times 0.98 \times 45.5 = 40.3$$

and from Table 12, for $p_y = 275$, $p_b = 247.2\,N/mm^2$

$$\therefore \quad M_b = S_x p_b = 2868 \times 0.2472 = 709.0\,kNm$$

The available buckling resistance moment M_b is therefore only 96% of the required value of 735.2 kNm. The reduced ultimate limit state load P which is available from $M_b = 709.0\,kNm$ is given by

$$\frac{PL}{2} = 709.0 + 404.8 \quad \text{i.e. } P = \frac{1113.8 \times 2}{6} = 371.3\,kN$$

Thus, the available capacity is only 97.7% of the required capacity of $P = 380\,kN$. At this stage, the designer would reconsider the loading to see whether a 2.3% reduction could be justified. If a reconsideration of the required minimum loading does not achieve a suitable reduction, then the flange plates would have to be increased from 250×12 to 250×15.

4.3.6 Effect of settlement

The first plastic hinges occur at the supports B and C at 1.115 times the unfactored loads whereas with the uniform section this factor was 1.503. Hence the non-uniform section has less capacity for settlement before a plastic hinge forms at the unfactored loading. It is acceptable to have a plastic hinge occurring at less than the unfactored loading provided that the loading is predominantly static and rarely occurring as is usual in building structures. Furthermore, strictly speaking, the reduced stiffness caused by the hinge has to be taken into account in the calculation of deflections at the serviceability limit state.

4.4 Alternative non-uniform section design for Example 4.2

An alternative design is to have the $457 \times 191 \times 67$ section only in the end spans AB and CD and to have a larger section in span BC. This would entail site bolted connections at B and C designed to enable a plastic hinge to form. It should be noted when designing such an arrangement that the bending moment will be the same on either side of the connection and that the plastic hinge will always tend to form in the weaker member.

Site connections can be useful on a congested site. Alternatively, the connections could be full-strength welded connections made in the fabrication shop or even on site. The sizing of the larger beam is easily done on the basis of stability calculations because the centre part of member BC is the design criterion with a constant bending moment along its length of 735.2 kNm. The beam has to be justified to clause 4.3.7 of BS 5950 and this can be done quickly from page 139 of Reference 4.1, thus:

Effective length $= 0.85 \times 3000 = 2500\,mm$ and $n = 1.0$

$M_b \geq 735.2\,kNm$

From the tables the lightest suitable section is a $610 \times 229 \times 113\,UB$ with

$$M_b = 807 - \frac{0.05(807 - 745)}{0.50} = 800\,kNm$$

4.5 Comparison of designs (Table 4.2)

Unless the beam depth or site congestion is critical the uniform section is likely to be the most economical solution. However, the fabrication shop and the site have to accommodate a beam 21 m long.

Table 4.2 Comparison of designs.

	Uniform section	Non-uniform sections	
		Plated beam	Different I beams
Beam depth (mm)	533	478	454 and 608
Beam weight (kg)	1932	1690	1821
Additional cost factors		Intermediate restraints, or flange plates 9 m long.	Full strength connections at B and C
		Flange plate welds	
		Additional Design and Drawing office time	

Reference

4.1 Steel Construction Institute. *Steelwork Design*: Guide to BS 5950: Part 1: 1990. Volume 1 – Section properties and member capacities. 3rd ed.

Chapter 5

Plastic Design of a Pitched Roof Portal Frame Building

5.1 Introduction

This chapter will illustrate some of the procedures described in the earlier chapters by giving an example of the design of the main frames of a complete portal frame building. The general arrangement of the building is shown in Fig. 5.1. The calculations cover the design of the primary structural members and the main connections.

5.2 Design of the main frames

The dimensions and loading of the main portal frames are shown in Fig. 5.2. They will be designed by plastic theory according to BS 5950: Part 1.

Dimensions
Span	22.5 m
Frame centres	6.0 m
Height to eaves	4.5 m
Height to underside of haunch	3.9 m assumed at this stage
Roof slope	6.0°
Typical purlin spacing on slope	1.6 m assumed at this stage
Typical purlin spacing in plan $= 1.6\cos 6°$	1.591 m

5.2.1 Downward loads (on plan area)

Dead load	Characteristic	Load factor	Ultimate
Sheets (PVC coated steel)	0.072		
Insulation			
(30 mm polyurethane foam)	0.010		
Steel liner tray	0.040		(continued on p. 156)

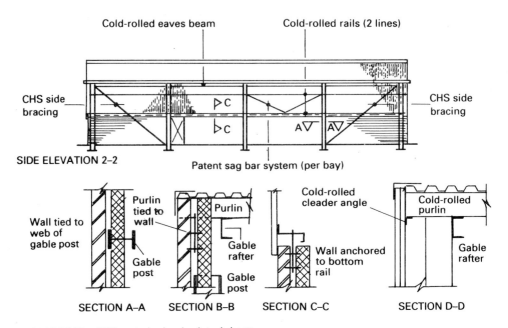

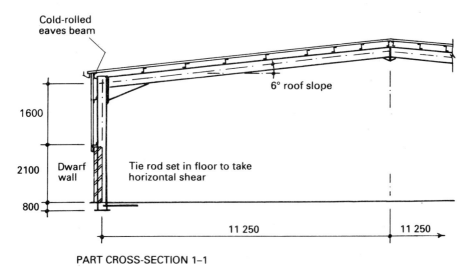

Fig. 5.1 General arrangement of portal frame building.

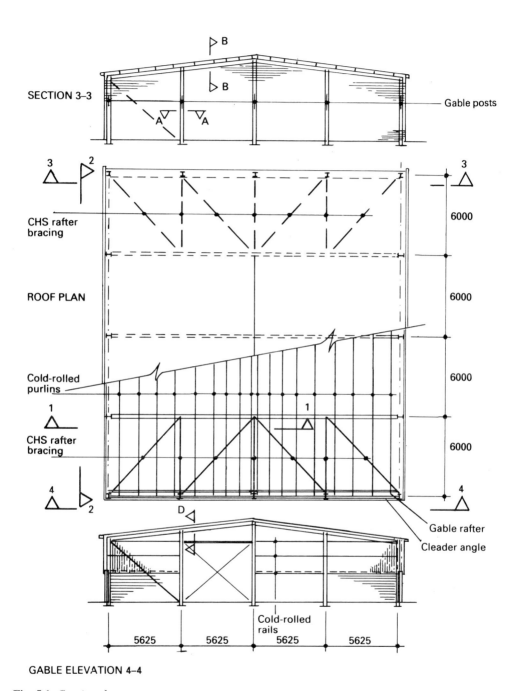

SECTION 3–3

Gable posts

3   2   3

CHS rafter
bracing   6000

ROOF PLAN   6000

Cold-rolled
purlins   6000

1   1

CHS rafter
bracing

4   6000
    2

D

                                                                    4

Gable rafter

Cleader angle

Cold-rolled
rails

5625   5625   5625   5625

GABLE ELEVATION 4–4

Fig. 5.1 *Continued.*

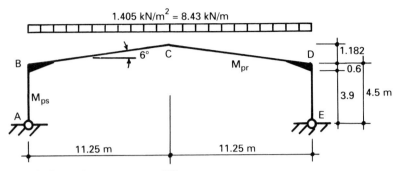

Fig. 5.2 General arrangement of frame.

Cold-rolled purlins	0.026	
Rafter self-weight	0.070	
Services	0.100	
Total dead load	0.318×1.4	$= 0.445 \, \text{kN/m}^2$
Live load (snow)	0.600×1.6	$= 0.960 \, \text{kN/m}^2$
Total vertical load	0.918	$= 1.405 \, \text{kN/m}^2$
Total load per frame	$1.405 \times 22.5 \times 6.0$	$= 189.7 \, \text{kN}$

5.2.2   Notional horizontal loads

The notional horizontal loads to be combined with the vertical load case according to clause 2.4.2.3 of BS 5950: Part 1 are:

$0.005 \times 189.7 \div 2 = 0.474 \, \text{kN}$ applied horizontally at each eave.

5.2.3   Wind load

Manchester area	$V = 45 \, \text{m/s}$
Topography factor	$S_1 = 1.0$
Height $= 5.68 \, \text{m} +$ purlins etc. $\approx 6 \, \text{m}$, condition 3, class B	$S_2 = 0.668$
Statistical factor	$S_3 = 1.0$

$$V_s = S_1 S_2 S_3 V = 45 \times 1.0 \times 0.668 \times 1.0 = 30.06 \, \text{m/s}$$

$$q = 0.613 \times 30.06^2 = 553 \, \text{N/m}^2$$

Note: Wind is not usually a criterion in the design of low-rise portal frames. It is therefore usual to design the frames on the basis of vertical load only (with the notional horizontal loads – see below) and then to check that the frame is safe under wind conditions.

5.2.4   Choice of the main frame members

There are a number of methods available for the choice of M_{ps} (for the stanchion) and M_{pr} (for the rafter), e.g.,

❑ Use computer-aided design.
❑ Use the work equation method
    (1)   with a preselected ratio of M_{pr} to M_{ps} (e.g. $M_{pr} = 0.6\,M_{ps}$);
    (2)   to find a minimum weight design semi-graphically.
❑ Use the approximate method for single-bay frames given in Appendix 5A.

Here we use a fourth possibility, the semi-graphical method, which was described in Section 2.7 and which is based on the reasoning shown in Fig. 5.3. This method takes advantage of the fact that a pinned base portal frame is only one degree redundant and it is convenient to work with the single unknown H.

    Under vertical load only, the free bending moment diagram is the same as that for the rafter treated as a simply supported beam in plan. The reactant bending moment diagram has the magnitude Hy at any point where y is the height of that point above the base. Superimposing these two bending moment distributions on a single diagram, where the shape of the frame is 'unwrapped' for simplicity gives the distribution shown shaded in Fig. 5.4.

    The part of the frame strengthened by the haunch can be easily identified on the diagram. Plastic hinges will not form within the haunch and we can extend the length L_h to ensure that no plastic hinge forms at the rafter end of the haunch. It follows that plastic hinges can only form at the symmetrical positions indicated on the diagram. To choose M_{pr} and M_{ps} we simply try different values of H until we obtain a reasonable fit with the available range of Universal Beam sections. For practical

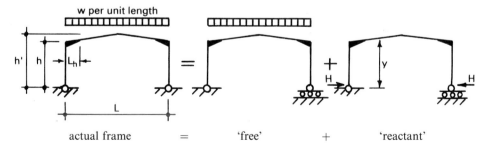

Fig. 5.3 Basis of the semi-graphical method.

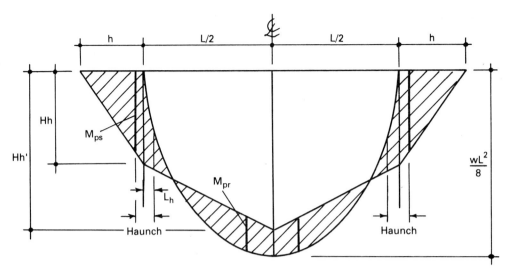

Fig. 5.4 Bending moment diagram using the semi-graphical method.

reasons, M_{pr} should not be allowed out of the following range and well-proportioned designs will usually be towards the middle of this range.

$$\frac{M_{ps}}{2.5} < M_{pr} < M_{ps}$$

For a symmetrical frame with symmetrical loading, it is only necessary to draw half of the above diagram. This is shown drawn to scale on graph paper in Fig. 5.5. The choice of sections proceeds as follows:

Try, as stanchion, $406 \times 140 \times 46$ UB   with $M_{ps} = 244$ kNm

Note: (1) Sections must be chosen which are classified as Class (1) 'plastic' according to BS 5950: Part 1, clause 3.5.2.
(2) The choice is usually based on the *lightest* section with the required full plastic moment M_p.
(3) The reduction of M_p due to axial thrust can usually be ignored in portal frames.

With a plastic hinge just below the haunch, this gives:

$$H = \frac{244}{3.9} = 62.56 \text{ kN}$$

The reactant bending moment at the eaves is then

$$4.5H = 281.5 \text{ kNm}$$

and at the apex is $5.682H = 355.5$ kNm

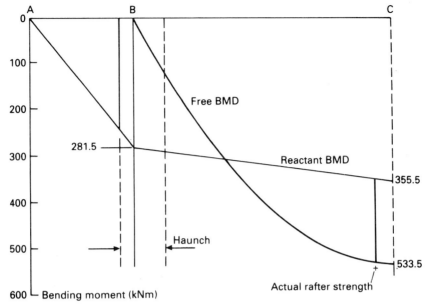

Fig. 5.5 Bending moment diagram for the design of the frame shown in Fig. 5.2.

so that the reactant line can be drawn on Fig. 5.5 as shown. The maximum bending moment in the rafter scales approximately 181 kNm.

We therefore choose, as rafter, $406 \times 140 \times 39$ UB with $M_{pr} = 198$ kNm.

As the rafter section is a little greater than the minimum required, the actual load factor at collapse will be greater than unity. This is necessary to allow for second-order effects. The actual load factor at collapse can be estimated quite accurately by first plotting the actual rafter strength on Fig. 5.5 at the rafter hinge position as shown. This corresponds to a free bending moment diagram ordinate of about 548 kNm compared with the value of 531 kNm in the free bending moment diagram as drawn at unit load factor. It follows that the load factor against plastic collapse is about $548/531 = 1.03$.

5.2.5 Influence of notional horizontal loads

The above choice of members has been carried out deliberately neglecting the influence of notional horizontal loads despite the fact that BS 5950: Part 1, clause 2.4.2.3 requires that they should be combined with the factored vertical loads. The advantage of neglecting these rather small additional loads is that the load case becomes symmetrical and it is only necessary to consider half of the bending moment diagram. However, neglecting the requirements of a British Standard should not be done without justification!

The notional horizontal forces are intended to take account of practical imperfections such as lack of verticality. However, lack of verticality only has a significant influence on the design of structures which are sensitive to second-order *sway* effects. It has already been argued in this publication that typical portal frames are not particularly sensitive to such sway effects but are much more influenced by second-order effects in the rafters. It would be more logical, therefore, to require consideration of notional forces which more directly aggravate the bending moments in the rafters.

This is the theoretical argument. The pragmatic argument is that the notional horizontal forces have little effect on the design. Figure 5.6 shows the design forces for the vertical load case taking into account the notional horizontal loads. These have their most significant effect on the eaves bending moment which now has free moment values of 4.27 kNm at B and zero at E whereas both were zero under vertical load alone. The free bending moment at the apex C increases from 533.5 to 535.6 kNm. In percentage terms, these changes are very small and their effect on the design is minimal.

The difference caused by the notional horizontal loads is too small to show on a figure such as Fig. 5.5. In terms of member selection, if the stanchion below the haunch is considered first, the notional horizontal loads are favourable at B and therefore consideration of the moments at D gives rise to precisely the same reactant bending moment diagram as before. The maximum bending moment in the rafter increases from 181 to 183 kNm; less than 1%. It is not considered that these effects are sufficiently significant to influence the design.

It is therefore suggested that, provided designers carry out an approximate check to demonstrate that the notional horizontal loads

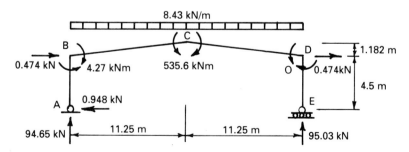

Fig. 5.6 Vertical load case with notional horizontal loads.

have an insignificant effect on the bending moments, it is not necessary to formally include them in the design calculations for low rise portal frames.

5.3   Check for second-order effects

Strictly speaking, second-order effects should be checked according to BS 5950: Part 1, clause 5.5.3.2 'sway stability'. However, this has been shown to be unsafe and the check will be carried out using the method given in Section 3.2.

Using Fig. 5.7 and the formula given in Section 3.2.3, the procedure is as follows:

At the ultimate load,

$$P_c = 94.84\,\text{kN} \quad \text{and} \quad H = 62.56\,\text{kN} \qquad \text{so that}$$

$$P_r = H\cos\theta + P_c\sin\theta = 72.13\,\text{kN}$$

$$I_c = 156.7 \times 10^6\,\text{mm}^4 \quad (406 \times 140 \times 46\,\text{UB})$$

$$I_r = 124.1 \times 10^6\,\text{mm}^4 \quad (406 \times 140 \times 39\,\text{UB})$$

$$L_r = 11.25\sec\theta = 11.312\,\text{m}$$

$$R = \frac{I_c I_r}{I_r h} = \frac{156.7 \times 10^6 \times 11.312}{124.1 \times 10^6 \times 4.5} = 3.174$$

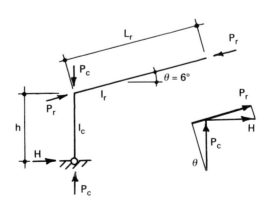

Fig. 5.7 Verification of sway stability.

The elastic critical load λ_{cr} is then given by

$$\lambda_{cr} = \frac{3EI_r}{L_r\left[\left(1 + \frac{1.2}{R}\right)P_c h + 0.3P_r L_r\right]}$$

$$= \frac{3 \times 205 \times 125.7 \times 10^6}{11\,312\left[\left(1 + \frac{1.2}{3.174}\right) \times 94.84 \times 4500 + 0.3 \times 72.13 \times 11\,312\right]}$$

$$= 8.10$$

The required load factor for plastic collapse (≥ 1.0) taking account of frame instability is

$$\lambda_p = \frac{0.9\lambda_{cr}}{\lambda_{cr} - 1} \qquad (4.6 < \lambda_{cr} < 10) = 1.027$$

As the actual value of λ_p is 1.03, the frame is safe with respect to second-order effects.

5.4   Length of haunch

The haunch should be sufficiently long to ensure that there is no plastic hinge at its rafter end. For this to be the case, the bending moment at the end of the haunch should be not greater than $0.87M_{pr} = 0.87 \times 198 = 172.3\,\text{kNm}$. Scaling off the bending moment diagram gives a minimum haunch length of 1.35 m.

Note: The factor of 0.87 arises from the consideration that the shape factor of a Universal Beam is approximately 1.15. If the rafter is to remain elastic, the limiting bending moment is $M_{pr}/1.15 = 0.87M_{pr}$.

5.5   Elastic-plastic computer analysis

This chapter is based on manual analysis. However, if an elastic-plastic computer analysis is available, it opens up some useful options when dealing with member stability. Such an analysis is given in Appendix 5B. This confirms the frame design that has been achieved so far. It may also be noted that it demonstrates that the last plastic hinge forms in the rafter at a load factor greater than unity and that there are no rafter hinges at the design loads. If this is known to the designer, it is not necessary to stabilise the rafter hinges for plastic rotation according to clause 5.3.5 of BS 5950: Part 1. Useful savings can then follow.

A single-bay frame, such as is being considered here, is only one degree redundant and therefore requires only two plastic hinges for *complete* collapse, one below the eaves haunch and one near the apex. In this case, the *first* hinge to form can be identified by an elastic analysis and the other hinge must therefore be the last.

It may be thought that an elastic analysis would be required in any case in order to determine the deflections. This is so as the designer needs this information even though BS 5950: Part 1, clause 2.5.1, and particularly Table 5, specifically excludes pitched roof portal frames from any deflection limitations.

5.6 Bracing of portal frame rafters

The outer flange of a portal frame is braced at intervals by the purlins and sheeting rails. Within the length of the rafter, the bending moment changes sign so that near the apex it is the compression flange which is restrained. Less helpfully, in the vicinity of the eaves it is the tension flange which is restrained. Sophisticated methods have been included in BS 5950: Part 1: Appendix G in order to take account of this periodic restraint to the tension flange. However, these are not easy to apply without the help of a computer and here we will confine ourselves to the simpler methods given in section 5.3 of the Standard.

Purlins provide positional restraint to the outer flange but not torsional restraint to the complete section. Where such torsional restraint is required, or where the purlin is attached to the tension flange and

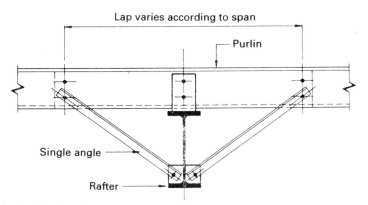

Fig. 5.8   Typical fly brace.

compression flange restraint is required, it is customary to provide 'fly braces', otherwise known as 'knee braces' as shown in Fig. 5.8. These important members are discussed in more detail in Section 3.3.5.3 of this book.

It is important to appreciate that a fly brace is only effective if the sizes of the frames and purlins are in proportion. In the authors' experience, this is not always the case. Thus a small purlin of say 125 mm depth can be used to stabilise a 356 mm deep rafter. However, it is no use at all to stabilise a 914 mm Universal Beam. Indeed, for large span frames which use hot-rolled sections at the heavier end of the available range, fly braces to cold-formed purlins are not usually adequate and a separate bracing system (usually using tubular members) may be required.

Another misconception that arises from time to time concerns the provision of heavy eaves beams as shown in Fig. 5.9. Some designers provide these as a matter of course but they serve no useful purpose. A light gauge steel eaves beam which supports the gutter and the edge of the roof sheeting as shown in Fig. 5.1 is all that is required.

A final point to note here is that, provided certain conditions are met, clause 5.5.3.5.2 allows a virtual lateral restraint to the *bottom* flange to be assumed at the point of contraflexure.

5.6.1   *Rafter stability based on manual analysis*

Ensuring the stability of the upper region of the rafter in the vicinity of the plastic hinge is one of the more troublesome aspects of the design of portal frames based on plastic theory. This is because:

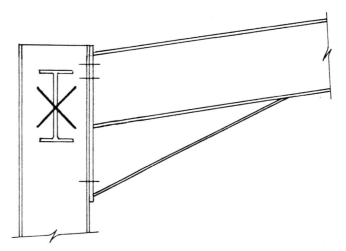

Fig. 5.9 Heavy eaves beams are *not* required.

❏ The plastic hinges in the rafters must be provided with torsional restraints both at the hinge position and within a specified distance adjacent to it.

❏ The presence of a plastic hinge means that the rafter must be designed to carry the full plastic moment of the cross-section. It is *not* possible to de-rate the rafter to a lower bending moment *unless* it can be demonstrated that no rafter plastic hinge is present at the required design load.

❏ The bending moment diagram in the vicinity of the hinge is almost uniform. This means that in calculations for lateral torsional buckling the factors *m* and *n* (see BS 5950: clause 4.3.7) are both equal to unity which is the worst possible condition.

In interpreting the above points, the crucial factor is that the design bending moment for rafter stability *must* include a plastic hinge in the upper region of the rafter *unless* the designer can prove that it is not there. Simple plastic theory offers no such proof. The bending moment for rafter design is therefore slightly different from that shown in Fig. 5.5 and the requirement is shown in Fig. 5.10 where the reactant line is drawn with the same free bending moment diagram as before but with the full plastic moment of the rafter ($M_{pr} = 198\,\text{kNm}$) at the rafter

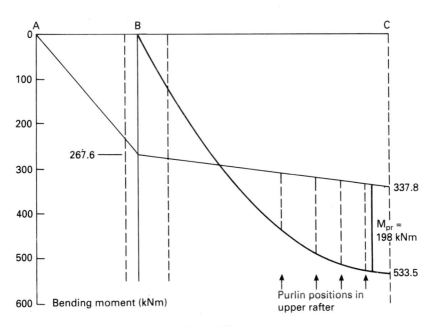

Fig. 5.10 Bending moment diagram for rafter stability.

hinge position. This second reactant line would normally be drawn on the original diagram. It is shown on a separate diagram here for clarity.

The required reactant line is simply drawn as follows. At the plastic hinge position, the free bending moment is 531 kNm and therefore the required ordinate of the reactant line is $531 - 198 = 333$ kNm. The height of the rafter at this point is $4.5 + 10.5 \tan 6° = 5.60$ m so that the horizontal thrust associated with this bending moment distribution is given by

$$5.60 H_r = 333 \text{ kNm}$$

i.e. $H_r = 59.46$ kN

This allows the reactant line to be drawn with an eaves ordinate of $4.5 H_r = 267.6$ kNm and an apex ordinate of $5.682 H_r = 337.8$ kNm. This diagram will be required to check other aspects of rafter stability once the purlin positions have been determined.

In general, the spacing of purlins in the vicinity of the rafter hinges should be determined using BS 5950: clause 5.3.5. This requires that the maximum distance L_m from the torsional restraint at the plastic hinge to an adjacent restraint should not exceed

$$L_m = \frac{38 r_y}{\left[\dfrac{f_c}{130} + \left(\dfrac{p_y}{275} \right)^2 \left(\dfrac{x}{36} \right)^2 \right]^{\frac{1}{2}}}$$

where f_c = the average compressive stress due to axial load (N/mm²)
  p_y = the design strength (N/mm²)
  r_y = radius of gyration about the minor axis (mm)
  x = torsional index of the rafter section

Here $f_c = \dfrac{P_r}{A_r} = \dfrac{72.13 \times 10^3}{4920} = 14.66 \text{ N/mm}^2$

$A_r = 49.2 \text{ cm}^2$
$r_y = 28.9 \text{ mm}$ section properties of
$x = 47.6$ $406 \times 140 \times 39$ UB

giving $L_m = 805$ mm.

This means that purlins with fly braces should be provided at the plastic hinge position and within a maximum distance of 805 mm adjacent to it. This is quite an onerous requirement and some designers make strenuous efforts to improve upon it.

5.6.2   An important improvement when the rafter hinge is the last to form

When it has been shown, by computer analysis or otherwise, that there is no plastic hinge in the rafter at the design load, it is advantageous to design the rafter using the actual bending moment distribution at the design load. This will usually allow the rafter to be designed to a lower bending moment than M_{pr} with a consequent relaxation of the requirements of BS 5950: section 4.

Here, the required bending moment distribution is the one shown in Fig. 5.5 where the reduced value of the maximum rafter bending moment is 181 kNm. Noting that the plastic section modulus of the rafter is equal to 718 cm^3, the corresponding design stress is

$$\frac{181 \times 10^6}{718 \times 10^3} = 252 \, \text{M/mm}^2$$

Reading down the third column of Table 11 of BS 5950: Part 1, or more simply using Fig. 5.11 which is plotted from Table 11, this is achieved at a value of $\lambda_{LT} = 44.1$ and therefore, for stability,

$$nuv \, \frac{L_E}{r_y} < 44.1 \quad \text{or} \quad L_E < \frac{44.1 r_y}{nuv}$$

Inserting the values appropriate to a $406 \times 140 \times 39$ UB rafter, namely

$n = 1.0$ (almost uniform bending moment)
$u = 0.859$ (tabulated buckling parameter)
$v = 1.0$ (conservative: a slightly better value may be obtained using Table 14 with $N = 0.5$)
$r_y = 28.9$ mm

gives the limiting value of the purlin spacing as:

$$L_m = 1484 \, \text{mm}$$

Obviously, when this approach is to be used, it may be advantageous not to be too ambitious when pruning the frame members to the minimum in order to achieve greater flexibility with the purlin spacing in this critical region. It may be noted that the more sophisticated computer-aided design packages on the market use the above approach. They carry out an elastic-plastic analysis to failure, noting the bending moment diagram at the design load and the plastic hinge history, and use these as the basis for member design. Thus they achieve the maximum economy with regard to the requirements of member stability.

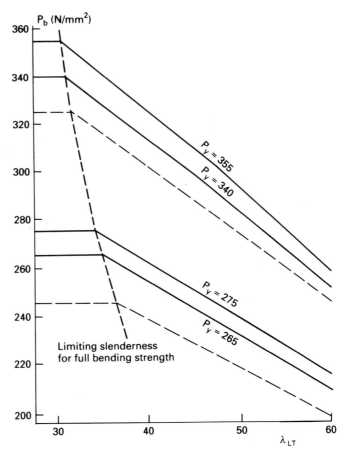

Fig. 5.11 Bending stresses as a function of slenderness from BS 5950, Part 1, Table 11.

5.6.3 *Layout of purlins along the rafter*

Having completed the stability checks for the upper part of the rafter, it is now necessary to lay out the purlin positions along the rafter. In doing this, the aim should be to minimise the number of purlins within the limitation of the spanning capability of the sheeting while arriving at a sensible arrangement at the eaves. Here, we will ignore the fact that we have used the results of an elastic-plastic computer analysis to justify a purlin spacing of 1484 mm and restrict ourselves to the maximum spacing of 805 mm justified by manual analysis. It may be noted here that, strictly speaking, the bending moment diagram is a series of straight lines between the purlin points rather than the parabola that we have been using for convenience up to this point. This means that, as we move the purlins about on the rafter, the plastic hinge position will follow one of the purlins. Here, the shape of the bending moment diagram indicates that the plastic hinge will form at the first purlin below the apex purlin. We should also note that the bending moment diagram is almost uniform near the apex and it is therefore advisable to stabilise the apex purlin as well as the first purlin from the apex.

A suitable purlin layout is then shown in Fig. 5.12.

Note: It is true that the last plastic hinge generally forms near the ridge and that it is industry practice *not* to stabilise purlins in this region because designers assume that the last hinge *always* forms there. To the best of the authors' knowledge, this has never been proved and therefore this book demonstrates what is considered to be current good practice *unless* the designer takes the trouble to confirm the location of the last hinge to form. Purlin stays may also be required in this region to cater for wind uplift.

The bending moments at the purlin positions near the apex can now be scaled off the graph or evaluated using

$$M_x = - H_r(4.5 + x \tan \theta) + \frac{8.43}{2} x(22.5 - x)$$

with $H_r = 59.46 \, \text{kN}$

This gives the following values:

Purlin number   6 $x = 7.27\,\text{m}$ $M = 153.7\,\text{kNm}$
                 7   $x = 8.86\,\text{m}$ $M = 186.5\,\text{kNm}$
                 8   $x = 9.66\,\text{m}$ $M = 194.9\,\text{kNm}$
                 9   $x = 10.45\,\text{m}$ $M = 197.9\,\text{kNm}$
                10   $x = 11.25\,\text{m}$ $M = 195.6\,\text{kNm}$

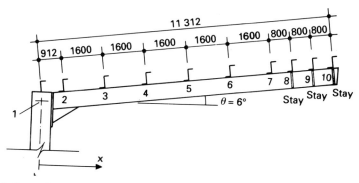

Fig. 5.12 Purlin layout to satisfy rafter stability.

At this point, the stability of each segment of the upper region of the rafter should be checked using section 4 of BS 5950. Here, in view of the preceding calculations, it is sufficient to verify that, when the purlin spacing has increased to 1.6 m between purlins 6 and 7, the bending moment has reduced sufficiently for this segment to be stable. Thus:

$$\beta = \frac{153.7}{186.5} = 0.824$$

$$\therefore \quad m = 0.910 \quad \text{(Table 18)}$$

$\therefore$ equivalent uniform moment $mM = 0.910 \times 186.5 = 169.7 \text{ kNm}$.

Therefore, from tables of buckling resistance moment with $n = 1$, or from section 4 of BS 5950, this segment is stable (permissible $L_E = 1.74$ m).

5.7 Stability of the haunch region

The next check to be made concerns the stability of the section of the haunched rafter between the eaves and the point of contraflexure. In this region, the purlins provide intermittent restraint to the *tension* flange but the compression flange is only restrained where stays are provided. However, clause 5.5.3.5.2 allows us to also assume an effective lateral restraint to the compression flange at the point of contraflexure (see Section 3.3.5.2 of this book).

The general arrangement of the haunch region is shown in Fig. 5.13. A stay at purlin number 2 is inevitable. Also inevitable is a stayed sheeting rail below the haunch in order to stabilise the plastic hinge in

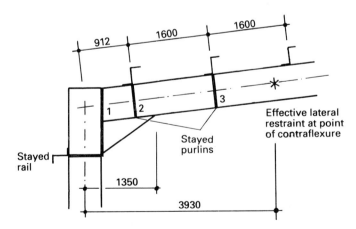

Fig. 5.13 Stability of the haunch region.

the stanchion at this position. This stayed rail will also serve to stabilise the inside of the haunch.

The requirements for the stability of this region are again given in clause 5.5.3.5.2 of BS 5950: Part 1. Figure 10 of the Standard shows the typical situations covered by this clause. There are two requirements, the first of which is that the length L_s between restraints to the compression flange should not exceed

$$L_s = \frac{K_1 r_y x}{(72x^2 - 10^4)^{\frac{1}{2}}} \quad \text{for Grade 43 steel}$$

where, here, the ratio of the depth of the haunch to the depth of the rafter may be assumed to be 2, giving $K_1 = 495$, so that

$$L_s = \frac{495 \times 28.9 \times 47.6}{(72 \times 47.6^2 - 10^4)^{\frac{1}{2}}} = 1740\,\text{mm}$$

which is satisfactory.

In addition, this region must be checked according to section 4 with L_E equal to the spacing of the *tension* flange restraints. For this calculation, the critical bending moment diagram is the original one shown in Fig. 5.5 with a plastic hinge in the column and $H = 62.56\,\text{kN}$. This gives the following bending moments at the critical locations:

Purlin number 1 $x = 0.20\,\text{m}$ (face of column) $M = -264.0\,\text{kNm}$
Purlin number 2 $x = 0.91\,\text{m}$ $M = -204.7\,\text{kNm}$
Purlin number 3 $x = 2.50\,\text{m}$ $M = -87.2\,\text{kNm}$

$$\therefore \quad \beta = \frac{-204.7}{-264.0} = 0.775$$

$$m = 0.886 \quad \text{(Table 18)}$$

$$\therefore \quad \text{equivalent uniform moment } mM = 0.886 \times 264.0$$
$$= 234.0\,\text{kNm}$$

In order to investigate whether this moment is acceptable, we need first to consider the section properties of the haunch.

5.7.1   Section properties of the haunch

The section properties of the haunch are not tabulated and must be calculated from first principles. Before we can do this, it is necessary to define the geometry of the haunch and this is done in Fig. 5.14.

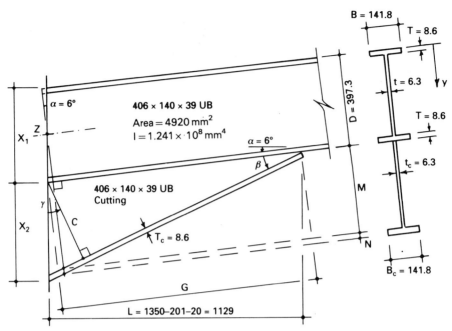

Fig. 5.14 Geometry of the haunch.

The middle flange of the haunch has little effect on the bending properties and, for simplicity, it can be ignored and the section assumed to be made up of three plates. However, the middle flange does have a beneficial effect on the minor axis radius of gyration and it should be included in the calculation of this property, even if it is omitted from the remainder of the calculation. Here, in order to demonstrate the procedure, all of the section properties will be calculated for the complete section shown on the right-hand side of Fig. 5.14.

The haunch will be fabricated from an offcut from the rafter section, namely a $406 \times 140 \times 39$ UB. For maximum economy, the full available depth is utilised and this defines the dimension C in Fig. 5.14 as the clear distance between the flanges, namely $397.3 - 2 \times 8.6 = 380.1$ mm. The length of the haunch from the column centre line was calculated in Section 5.4 as 1350 mm. Deducting half the depth of the column and the thickness of the end plate gives the plan length L as $1350 - 201 - 20 = 1129$ mm. The remaining dimensions in Fig. 5.14 are then simply a matter of geometry. Thus

$$X_1 = \frac{D}{\cos \alpha} = \frac{397.3}{\cos 6°} = 399.49 \text{ mm}$$

$$\beta = \tan^{-1}\left(\frac{C}{G}\right) = \tan^{-1}\left(\frac{380.1}{1135.2}\right) = 19.56°$$

$$\gamma = \alpha + \beta = 6 + 19.56 = 25.56°$$

$$X_2 = \frac{(380.1 + 8.6)}{\cos\gamma} = 430.87\,\text{mm}$$

$$M = \left(g - \frac{D}{2}\tan\alpha\right)\tan\beta = (1135.2 - 198.65\tan 6°)\tan 19.56°$$

$$= 395.96\,\text{mm}$$

$$N = \frac{T_c}{\cos\beta} = \frac{8.6}{\cos 19.56°} = 9.13\,\text{mm}$$

Bearing in mind that Appendix B3 of the Standard requires that the bending strength p_b is determined using the properties of the section at the point of maximum moment, the required properties are determined on the normal to the centre line of the rafter through the point Z.

For elastic bending about the major axis,

$$A = 4920 + 395.96 \times 6.3 + 141.8 \times 9.13$$

$$= 4920 + 2494.6 + 1294.6 = 8709.2\,\text{mm}^2$$

$$Ay_G = 4920\left(\frac{397.3}{2}\right) + 2494.6 \times 595.28 + 1294.6 \times 797.83$$

$$= 3\,494\,900\,\text{mm}^3$$

$$\therefore \quad y_G = \frac{3\,494\,900}{8\,709.2} = 401.3\,\text{mm}$$

For elastic bending about the minor axis,

$$I_x = 12\,410 + 49.2 \times 20.265^2$$

$$+ \frac{0.63 \times 39.596^3}{12} + 24.946 \times 19.398^2$$

$$+ \frac{141.8 \times 0.913^3}{12} + 12.942 \times 39.652^2$$

$$= 12\,410 + 20\,205.0 + 3\,259.2 + 9\,386.7 + 9.0 + 20\,348.5$$

$$= 65\,618.4\,\text{cm}^4$$

$$I_y = \frac{1}{12} \, [2 \times 8.6 \times 141.8^3 + 9.13 \times 141.8^3$$

$$+ (380.1 + 395.96)6.3^3]$$

$$= 6\,272\,200 \, \text{mm}^4$$

$$\therefore \quad r_y = \sqrt{\frac{I_y}{A}} = \sqrt{\frac{6\,272\,200}{8709.2}} = 26.84 \, \text{mm}$$

and for plastic bending about the major axis,

$$\frac{A}{2} = 4354.6 \, \text{mm}^2$$

$$\therefore \quad y_p = 397.3 - \frac{(4920 - 4354.6)}{2 \times 141.8} = 395.3 \, \text{mm}$$

$$\therefore \quad S = 49.2(39.53 - 19.865) + 24.946(0.20 - 19.798)$$

$$+ 12.942(0.20 + 39.596 + 0.456)$$

$$= 1987.4 \, \text{cm}^3$$

The design moment M_b then follows conservatively using $u = v = 1$:

$$\lambda_{LT} = nuv \, \frac{L_E}{r_y} = 1.0 \times 1.0 \times 1.0 \times \frac{710}{26.84} = 26.4$$

$$p_b = 275 \, \text{N/mm}^2 \quad \text{(Table 11)}$$

$$\therefore \quad M_b = 0.275 \times 1987 = 546 \, \text{kNm}$$

As this is greater than the equivalent uniform moment $mM = 234.0 \, \text{kNm}$, the first segment of the rafter is safe.

Note: Provided that the individual components making up the haunch have been chosen to be compact to clause 3.5.4 of BS 5950, it is *not* necessary to check the full depth of the web with regard to section classification. The middle flange of the haunch acts as a web stiffener and divides the web into two parts so that local buckling over the full depth of the web is not possible.

Unless purlin number 3 is stayed, the second segment of the rafter extends from purlin number 2 to the point of contraflexure, a distance of 3.04 m. Here, the section is uniform so that, in the calculation of L_s, $K_1 = 620$. This gives:

$$L_s = \frac{620 \times 28.9 \times 47.6}{(72 \times 47.6^2 - 10^4)^{\frac{1}{2}}} = 2180 \, \text{mm}$$

It follows that a stay is also required at purlin number 3. It is then obvious that the rafter between purlins 2 and 3 is stable so that this completes the verification of rafter stability.

5.8 Stanchion stability

The first requirement for stanchion stability has already been discussed in connection with Fig. 5.13. There is a plastic hinge in the stanchion below the haunch and this requires a torsional restraint. This is provided in the form of a sheeting rail and fly brace.

It is also necessary to provide a second torsional restraint within a specified distance of this plastic hinge. As in the case of the rafter hinge, this distance may be calculated conservatively as L_m in clause 5.3.5. However, here there is a bending moment gradient so the more favourable method given in Section 3.3.3 is also applicable. Nevertheless, the BS 5950 method suffices and is used to demonstrate that the stanchion is stable:

$$L_m = \frac{38 r_y}{\left(\frac{f_c}{130} + \left(\frac{p_y}{275} \right)^2 \left(\frac{x}{36} \right)^2 \right)^{\frac{1}{2}}}$$

where $f_c = \dfrac{94.84 \times 10^3}{5900} = 16.07\,\text{N/mm}^2$

$p_y = 275\,\text{N/mm}^2$

$x = 38.8 \quad (406 \times 140 \times 46\,\text{UB})$

$r_y = 30.3\,\text{mm}$

i.e. $L_m = \dfrac{38.8 \times 30.3}{\left(\dfrac{16.07}{130} + \left(\dfrac{38.8}{36} \right)^2 \right)^{\frac{1}{2}}} = 1015\,\text{mm}$

An additional braced rail is therefore required 1.0 m below the bottom of the haunch.

It is now necessary to check the 2.9 m length of stanchion below this second rail. As the bending moment diagram is linear, the bending moment at the position of the rail is $244.0 \times 2.9/3.9 = 181.4\,\text{kNm}$.

$\beta = 0$ (pinned at base)

$\therefore \quad m = 0.57$ (Table 18)

$\therefore$ equivalent uniform moment $= mM = 0.57 \times 181.4 = 103.4\,\text{kNm}$

$$\lambda_{LT} = nuv\lambda = 1.0 \times 0.87 \times 1.0 \times \frac{2900}{30.3} = 83.3$$

$\therefore$ $p_b = 157.7\,\text{N/mm}^2$ (Table 11)

$\therefore$ $M_b = S_x p_b = 889 \times 157.7 \times 10^{-3} = 140.1\,\text{kNm}$

$$p_c = 148\,\text{N/mm}^2 \quad \left(\text{Table 27(b) with } \lambda = \frac{2900}{30.3} = 95.7\right)$$

Therefore, the overall buckling check to clause 4.8.3.2 gives

$$\frac{F}{A_g p_c} + \frac{mM_x}{M_b} = \frac{94.84}{5900 \times 0.148} + \frac{103.4}{140.1} = 0.85$$

Therefore the lower section of the stanchion is satisfactory.

Having chosen the member sizes for the frame and verified their stability, we now turn our attention to the design of the connections. The procedure for connection design which follows is both simple and conventional. A more rigorous approach is given in Reference 5.1.

5.9   Design of the eaves connection

A flush end-plate connection will be used with the tentative design shown in Fig. 5.15. The following details in this design require to be checked:

Capacity of the bolt group to resist moment (5.9.1)
Capacity of the bolt group to resist shear (5.9.2)
Shear capacity of the column web (5.9.3)
Strength of the end-plate in bending (5.9.4)
Capacity of the flange of the stanchion to resist tensile forces (5.9.5)
Bearing and buckling of the stanchion web in compression (5.9.6)
Strength of the welds connecting the haunch to the end-plate (5.9.7)
Capacity of the web of the stanchion to resist tensile forces (5.9.8)
Capacity of the web of the rafter to resist tensile forces (5.9.8)

The design bending moment is the bending moment at the face of the column which has already been determined. The design shear force is half of the total vertical load. The bolts are M20 grade 8.8 bolts in clearance holes.

It should be noted that, for frames of larger span with deeper sections, the alternative connection design with an extended end-plate,

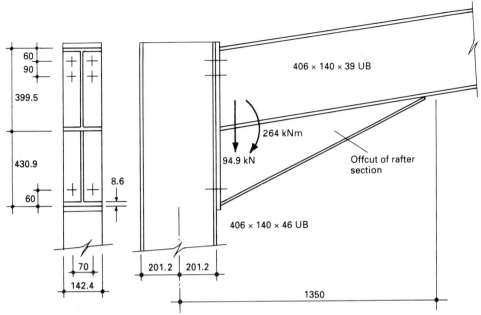

Fig. 5.15 Eaves connection.

as shown in Fig. 5.16, may be advantageous. However, this extended end-plate cannot be used in a number of frequently occurring situations such as when boundary wall gutters or curved eaves sheets are used.

5.9.1   Capacity of the tensile bolt group

It is assumed that the centre of rotation of the bolt group is at the centre of the compression flange. It is assumed that the top two rows of bolts reach their design strength and that other rows of bolts reach a reduced strength based on a linear strain distribution.

All shear is transmitted through the bottom row of bolts of the compression flange.

The tensile capacity of a single bolt $= 110\,\text{kN}$

$\therefore$ moment capacity of connection $= 2 \times 110\,(0.7661 + 0.6761)$

$$= 317.3\,\text{kNm}$$

Therefore the tensile bolt group is satisfactory.

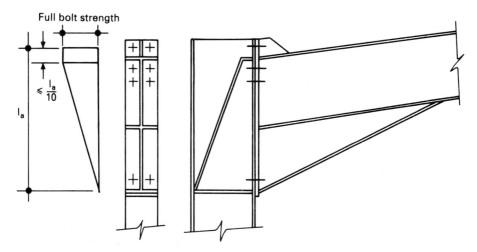

Fig. 5.16 Typical extended end plate connection.

When there are more than two rows of tension bolts, as shown in Fig. 5.16, it is common practice to assume that only the rows of bolts within the top 10% of the lever arm reach their full design strength. The other rows of bolts are assumed to reach a reduced strength based on a linear strain distribution extending from zero at the centre of rotation to full strength at 90% of the lever arm.

5.9.2 Capacity of the shear bolt group

Capacity of two M20 grade 8.8 bolts in single shear (tensile area $A_s = 245\,\text{mm}^2$)

$$= 2 \times 245 \times 0.375 = 184\,\text{kN} \quad \text{(clauses 6.3.1 and 6.3.2)}$$

Therefore the shear bolt group is satisfactory.

5.9.3 Shear capacity of the column web

The combined action of the tensile force in the upper bolt group and the compressive bearing at the lower flange of the haunch causes a considerable shear force in the haunch region of the stanchion web. This frequently requires a shear stiffener to be incorporated in this region as shown in Fig. 5.17. However, it is often more economic to increase the depth of the haunch until this stiffener becomes unnecessary.

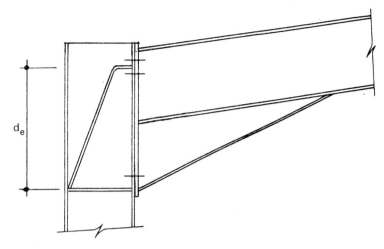

Fig. 5.17 Web shear stiffener.

The lever arm for this shear force is $d_e = 826.1 - 105 = 721.1\,\text{mm}$. Therefore the design shear force is

$$F_v = \frac{264 \times 10^3}{721.1} = 366.1\,\text{kN}$$

and the shear strength is (clause 4.2.3)

$$P_v = 0.6tDp_y$$

$$= 0.6 \times 6.9 \times 402.3 \times 0.275 = 458\,\text{kN}$$

Therefore the web is satisfactory in shear *without* a stiffener.

If a shear stiffener is required, the arrangement shown in Fig. 5.17 is generally advantageous, combining a tension stiffener (see Fig. 5.19) and a shear stiffener in one member. Page 219 of Reference 5.2 gives guidance on the design of this arrangement. The principle is to design the shear stiffener as a diagonal bracing member carrying the excess shear force which cannot be accommodated by the web of the stanchion.

5.9.4   *Thickness of the end-plate*

There is no single agreed method for checking the thickness of the end-plate. However, a good starting point is to make the end-plate thickness at least equal to the bolt diameter. A reasonable model for carrying out an approximate numerical check is as shown in Fig. 5.18. Then:

$$t_p = \sqrt{\frac{F_t m}{p_y L_e}}$$

where F_t = force carried by four bolts
 = $4P_b$
 L_e = length of the end-plate along which the bolt load is carried
 = lesser of $c + 3.5\,m$ or $7\,m$
 m = length across the end-plate over which double curvature occurs
 p_y = yield stress of the end-plate
 = $275\,\text{N/mm}^2$ for the plate material ($265\,\text{N/mm}^2$ when $t > 16\,\text{mm}$)

Here, therefore

$$F_t = \frac{264}{0.7211} = 366\,\text{kN}$$

$$a = \frac{142.4 - 70}{2} = 36.2\,\text{mm}$$

$$m = \frac{70}{2} - \frac{6.3}{2} - 6 = 25.85\,\text{mm}$$

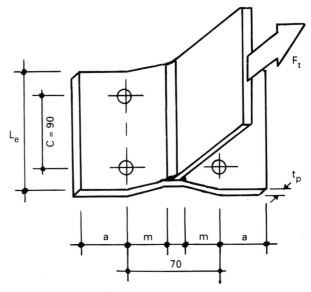

Fig. 5.18   Plastic collapse of end-plate.

$$7m = 180.95 \, \text{mm}; \qquad 3.5m + 90 = 180.5 \, \text{mm}$$

$$\text{so that} \quad t_p = \sqrt{\frac{366 \times 25.85}{0.275 \times 180.5}} = 13.8 \, \text{mm}$$

A 15 mm thick end-plate will, therefore, suffice.

5.9.5 *Strengthening of the flange of the stanchion in tension*

The thickness of the stanchion flange (11.2 mm) is less than that of the end-plate (15 mm). It may be assumed, therefore, that the flange of the stanchion requires strengthening with a stiffener in order to prevent it from being overstressed by the tensile bolt force. It is usual to make this stiffener of thickness similar to the flange of the rafter (8.6 mm) and to carry it through at least the centre line of the stanchion as shown in Fig. 5.19.

It may be noted that the web shear stiffener shown in Fig. 5.17 also combines the function of a tension stiffener. The welds between this stiffener and the stanchion should be fillet welds of throat thickness comparable with the elements being connected. Here an 8 mm thick plate is used with 6 mm fillet welds to the flanges and 6 mm welds to the web.

If the reader requires a more rigorous check of the adequacy of the flange of the stanchion, a suitable design expression is given on page 218 of Reference 5.2. The calculation proceeds as follows:

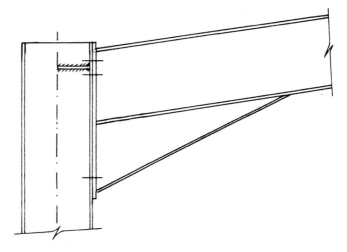

Fig. 5.19 Flange stiffener (tension).

Stiffening is required if

$$F_t \geq T_c^2 \left[\frac{c + w + w^*}{m} + \left(\frac{1}{w} + \frac{1}{w^*} \right)(m + n) \right] p_{yc}$$

where, in addition to quantities defined above

T_c = thickness of column flange = 11.2 mm (406 × 140 × 46 UB)

$$m = \frac{70}{2} - \frac{6.9}{2} - \frac{20.4}{2} = 21.35 \, \text{mm}$$

$$n = \frac{B - A}{2} = \frac{142.4 - 70}{2} = 36.2 \, \text{mm}$$

A = centres of holes = 70 mm

B = width of holes = 142.4 mm

$$w = \sqrt{m(m + n)} = \sqrt{21.35(21.35 + 36.2)} = 35.05 \, \text{mm}$$

$w^* = 70 \, \text{mm} < 2w = 70 \, \text{mm}$

This results in a right-hand side value of:

$$11.2^2 \left[\frac{90 + 35.05 + 70}{21.35} + \left(\frac{1}{35.05} + \frac{1}{70} \right)(21.35 + 36.2) \right] 0.275$$

$$= 400.2 \, \text{kN}$$

which is greater than $F_t = 366 \, \text{kN}$ so that the more rigorous check indicates that a tension stiffener is not, in fact, required.

Note: With larger rafter sections, it is usual to proportion the tension stiffeners on the basis of an estimate of the proportion of the bolt load carried by the stiffeners. This is based on the bolt horizontal and vertical cross centres, thus:

Load in the pair of stiffeners $= 336.1 \times \dfrac{70}{70 + 90} = 160.2 \, \text{kN}$

∴ $366.1 - 160.2 = 205.9 \, \text{kN}$ goes into the stanchion web

Using two No. 60 × 8 stiffeners and allowing for a corner mitre gives a stress of

$$\frac{160.2 \times 10^3}{2 \times 50 \times 8} = 200.3 \, \text{N/mm}^2$$

The welds to the stanchion web can be sized in accordance with Fig. 15 of BS 5950: Part 1: hence use 6 mm fillet welds.

The length of the stiffeners can be twice their width $= 120$ mm with 6 mm fillet welds to the web of the stanchion.

5.9.6   Bearing and buckling of the stanchion web in compresssion

A similar check is required at the bottom of the haunch where the compressive force from the lower flange of the haunch bears on the flange of the stanchion. This compressive force is tending to cause bearing and buckling of the web of the stanchion and a check of these conditions is essential.

The horizontal component of the compressive force in the flange of the haunch is

$$\frac{264 \times 10^3}{721.1} = 366.1 \text{ kN}$$

The relevant clauses in BS 5950: Part 1 are 4.5.2 for buckling and 4.5.3 for bearing. Thus, the buckling resistance of the unstiffened web is

$$P_w = (b_1 + n_1)tp_c$$

where b_1 is the stiff bearing length (equal here to the thickness of the flange of the haunch)

n_1 is the length obtained by dispersion at 45° through half of the depth of the section

t is the web thickness

p_c is the compressive strength assuming a web slenderness of $2.5d/t$ and using the appropriate strut table 27(c).

Thus here, $b_1 = 8.6$ mm

$$n_1 = 402.3 \text{ mm}$$

$$t = 6.9 \text{ mm}$$

$$\lambda_w = \frac{2.5 \times 359.7}{6.9} = 130.3$$

$$\therefore \quad p_c = 86 \text{ N/mm}^2$$

$$P_w = (8.6 + 402.3) \times 6.9 \times 86 \times 10^{-3} = 243.8 \text{ kN}$$

As this is less than the applied compressive force of 366.1 kN, a stiffener is required.

Similarly, the bearing resistance is $P_{wb} = (b_1 + n_2)tp_{yw}$ where, in addition to the quantities defined above

n_2 is the length obtained by dispersion through the flange to the flange-to-web connection at a slope of $1:2.5$ to the plane of the flange

p_{yw} is the design strength of the web

Thus here $n_2 = 2 \times 2.5 \times (11.2 + 10.2) = 107.0\,\text{mm}$

$$p_{yw} = 275\,\text{N}/\text{mm}^2$$

$$\therefore \quad P_{wb} = (8.6 + 107.0) \times 6.9 \times 275 \times 10^{-3} = 219.4\,\text{kN}$$

and the stanchion web also fails in bearing.

As we are concerned with buckling of the stanchion web as a strut of length equal to the depth of the section, it is necessary to provide a stiffener extending over the full depth of the section as shown in Fig. 5.20. This stiffener should be designed for both buckling and bearing.

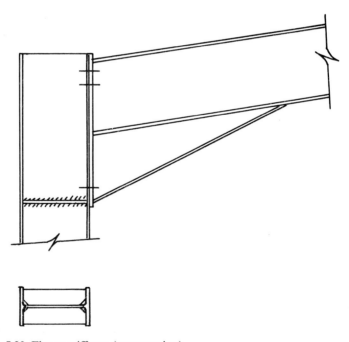

Fig. 5.20 Flange stiffener (compression).

The design of load-carrying stiffeners is considered in BS 5950: Part 1: clause 4.5.4. The first requirement is a bearing check which necessitates a stiffener whose area A *in contact with the flange* is given by

$$A > \frac{0.8F_x}{p_{ys}}$$

where F_x = external load

p_{ys} = design strength of stiffener

Thus here, $A = \dfrac{0.8 \times 366.1}{0.275} = 1065\,\text{mm}^2$

Bearing in mind that part of the stiffener needs to be cut away in order to clear the root radius, try two No. 60×12 plate stiffeners, one on either side of the stanchion web. These will be fully effective according to clause 4.5.1.2. The buckling resistance is then given in clause 4.5.1.5 with the effective cross-section shown in Fig. 5.21.

$$A_E = 276 \times 6.9 + 120 \times 12 = 3344.4\,\text{mm}^2$$

$$I_y = \frac{12 \times 126.9^3}{12} + \frac{(276 - 12) \times 6.9^3}{12}$$

$$= 2050.8 \times 10^3\,\text{mm}^4$$

$$\therefore \quad r_y = \sqrt{\frac{2050.8 \times 10^3}{3344.4}} = 24.76\,\text{mm}$$

$$L = D_c - 2T_c = 402.3 - 2 \times 11.2 = 379.9\,\text{mm}$$

$$L_E = 0.7L = 0.7 \times 379.9 = 265.9\,\text{mm}$$

$$\lambda = \frac{L_E}{r_y} = \frac{265.9}{24.76} = 10.7$$

For $p_y = 275\,\text{N/mm}^2$, $p_c = 275\,\text{N/mm}^2$ (Table 27(c))

$$\therefore \quad P_x = p_c A_E = 275 \times 3344.4 \times 10^{-3} = 919.7\,\text{kN}$$

which is greater than the required value of 366.1 kN and therefore amply safe.

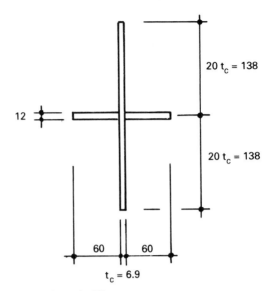

20 t_c = 138

12

20 t_c = 138

60   60

t_c = 6.9

Fig. 5.21 Cross-section of stiffener.

5.9.7   *Strength of the welds connecting the haunched rafter to the end-plate*

The strength of the welds connecting the rafter and haunch to the end-plate is critical. Fillet welds rather than butt welds are generally used in order to reduce the risk of delamination within the end-plate. Fillet welds are also preferred because butt welds shrink on cooling and the resulting distortions are generally disadvantageous and may reduce the effective lever arm of the connection. Butt welds also have more severe inspection requirements.

The stress distribution at the upper end of this connection is complex and empirical design is appropriate. A simple rule, which has been justified by tests, is to make the combined throat thickness of the welds at least equal to the thickness of the plate element being connected. Therefore:

$$\text{Leg length of flange weld} = \frac{T_b}{\sqrt{2}} = \frac{8.6}{\sqrt{2}} = 6.1\,\text{mm} \quad \text{top and bottom}$$

$$\text{Leg length of web weld} = \frac{t_b}{\sqrt{2}} = \frac{6.3}{\sqrt{2}} = 4.5\,\text{mm} \quad \text{both sides}$$

Therefore adopt 8 mm fillet welds for the flanges and 6 mm fillet welds for the web. The heavier weld *must* be continued down the web on the tension side for at least 50 mm in order to prevent tension cracking in the region of high stress concentration.

Note: Here also it is possible to adopt a more formal approach to weld sizing based on Fig. 15 of BS 5950: Part 1 and the bolt cross centres as described in the note to Section 5.9.5.

5.9.8   Rafter and stanchion webs in tension

Using a procedure similar to that described in Section 5.9.5:

$$\text{The rafter web load} = \frac{366.1}{2}\left(1 + \frac{55}{55 + 35}\right) = 294.9 \text{ kN}$$

$$\text{Hence the web stress} = \frac{294.9 \times 10^3}{211 \times 6.3} = 222 \text{ N/mm}^2$$

where the effective web length $= 60 + 90 + 1.75 \times 35 = 211$ mm.

From Section 5.9.5, the stanchion web load is 205.9 kN.

$$\text{Hence the web stress} = \frac{205.9 \times 10^3}{211 \times 6.9} = 142 \text{ N/mm}^2$$

5.10   Design of the ridge connection

The design bending moment at the ridge can be scaled from Fig. 5.10 and is 197 kNm.

Note: The bending moments are usually almost constant in the region of the apex and there is little loss of economy if the connection is designed for the full plastic moment of the rafter (here $= 198$ kNm).

The design of the ridge connection is much easier than that of the eaves connection. The haunch is only necessary in order to increase the lever arm of the bolts in order to obtain a sensibly sized bolt group in the tension region. Shear is nominally zero under symmetrical vertical load though two bolts are invariably included on the compression side. These serve to carry the shear force in an unsymmetrical load condition and to cater for any bending moment reversal under wind uplift.

With the dimensions shown in Fig. 5.22, the lever arm of the bottom tension bolts from the centre of the compression flange is $400 + 60 + 90 - 15 - 4.3 = 530.7$ mm. The lever arm of the upper tension bolts is therefore $530.7 - 90 = 440.7$ mm.

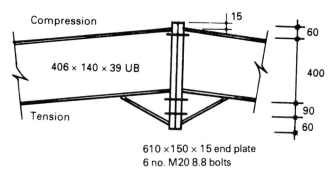

Fig. 5.22 Ridge connection.

The tensile capacity of a single M20 8.8 bolt is 110 kN and, bearing in mind that the 10% rule described in Section 5.9.1 is not satisfied, it is conservative to assume a force in the upper bolts based on a linear distribution of strain. The moment capacity of the connection is therefore:

$$2 \times 110 \left(0.5307 + \frac{0.4407^2}{0.5307} \right) = 197.3 \, \text{kNm}$$

The tensile bolt group is therefore just satisfactory.

The calculation of the thickness of the end-plate is as before (Section 5.9.4 and Fig. 5.18) so that

$$t_p = \sqrt{\frac{F_t m}{p_y L_e}} = \sqrt{\frac{4 \times 110 \times 25.85}{0.275 \times 180.5}} = 15.1 \, \text{mm}$$

A 15 mm end-plate will therefore suffice, bearing in mind that the rafter and cutting flanges will give rise to a stronger yield line mechanism than the one assumed.

The empirical rules for the weld sizes are also the same as before so that:

> leg length of flange weld $= 8$ mm (carried 50 mm up the web from the tension flange)
>
> leg length of web weld $= 6$ mm

The inclined flange of the ridge 'haunch' applies a tensile force to the bottom flange of the rafter. If there are no rafter stiffeners at the junction, the full flange width weld length will not be fully effective. Assuming that the stiff weld length is based on a $1 : 2.5$ bevel off the root radius of the rafter section, the effective weld length is:

> web thickness $+ 1.615 \times$ root radius $+ 5 \times$ flange thickness
>
> $= 6.3 + 1.615 \times 10.2 + 5 \times 8.6 = 66$ mm

The inclined flange force resolves into a shear force parallel to the flange of the rafter and a tensile force perpendicular to it. Sixty-six millimetres of the weld takes the tensile force and the remaining 130 mm is available to carry the shear force. Assuming that the load from one bolt goes into the inclined flange, then the shear force is (say) $0.85 \times 110 = 93.5 \, kN = 93.5/130 = 0.719 \, kN/mm$ and the tensile force is (say) $0.55 \times 110 = 60.5 \, kN = 60.5/66 = 0.917 \, kN/mm$. The resultant force in the weld is therefore:

$$\sqrt{0.719^2 + 0.917^2} = 1.17 \, kN/mm$$

Therefore use an 8 mm fillet weld $= 1.21 \, kN/mm$.

5.11   Design of the base plate

As the base is nominally pinned, and as the horizontal shear has been taken out by tie rods cast into the floor slab (see Fig. 5.1), the base has to carry vertical load only.

The design of base plates is covered in BS 5950: Part 1: clause 4.13.2. However, it will be found that, if the formula given in this clause is applied to the compact base plates usually used for nominally pinned connections, the minimum thickness given is impractically small. Once again, therefore, the design is empirical.

A practical design approach is to make the base plate thickness at least equal to the thickness of the flange of the stanchion (11.2 mm). The design shown in Fig. 5.23 can therefore be adopted without calculation.

Note: A four-bolt connection is shown. Two bolts on the centre line of the stanchion would suffice and such a detail is often used. However, four bolts makes for easier erection and may in fact be required by the Health and Safety Executive if the erection sequence requires the stanchions to act temporarily as free standing cantilevers. A four-bolt detail may also be necessary if the frame is to be designed for boundary fire conditions.

5.12   Check under wind loads

Unless the frame is unusually high, or unless other unusual conditions prevail, wind loads are not usually critical in the UK in the design of pitched roof portal frames for the ultimate limit state. In other parts of

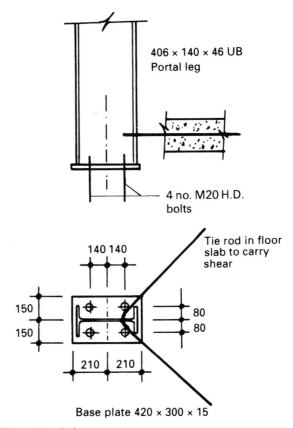

406 × 140 × 46 UB
Portal leg

4 no. M20 H.D.
bolts

Tie rod in floor
slab to carry
shear

140 140

150
150

80
80

210 | 210

Base plate 420 × 300 × 15

Fig. 5.23 Base plate design.

the world, where snow is not the dominant load case, the reverse may be true and the wind load cases become dominant.

However, wind load deflections may be critical, for instance where masonry walls are used (see also Section 1.9).

5.12.1   Wind load cases

The dynamic pressure has been calculated previously as $q = 553 \text{ N/mm}^2$.

The external pressure coefficients C_{pe} follow from CP3: Chapter V: Part 2: Tables 7 and 8 and are shown in Fig. 5.24. The internal pressure coefficients $C_{pi} = +0.2$ or -0.3 are for normal permeability as given in Appendix E of CP3.

It is usually sufficient to consider two transverse wind load cases A and B below. Note, however, that longitudinal wind can sometimes

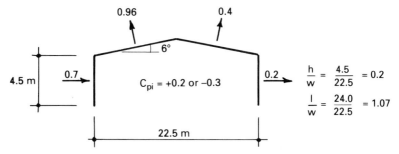

Fig. 5.24 Pressure coefficients.

produce the critical case for wind uplift on the rafters and determine the
requirements for rafter stability:

Case A: Full vertical and wind loads with all partial load factors $= 1.2$.
Internal suction is considered in order to give rise to the
maximum side load on the windward stanchion.

Case B: Dead load with a partial load factor of 1.0 and wind load with
a partial load factor of 1.4. Internal pressure is considered in
order to give rise to maximum uplift on the rafters.

The resultant pressure coefficients $(C_{pe} - C_{pi})$ for these two cases are
shown in Fig. 5.25.
    The uniformly distributed loads for these two cases can be calculated
as follows:

Case A: vertical load $= 0.918 \times 1.2 \times 6 = 6.610 \, \text{kN/m}$
          wind load $AB = 1.0 \times 0.553 \times 1.2 \times 6 = 3.982 \, \text{kN/m}$
          $BC = 0.66 \times 0.553 \times 1.2 \times 6 = 2.628 \, \text{kN/m}$
          $CD = 0.1 \times 0.553 \times 1.2 \times 6 = 0.398 \, \text{kN/m}$
          $DE = 0.1 \times 0.553 \times 1.2 \times 6 = 0.398 \, \text{kN/m}$

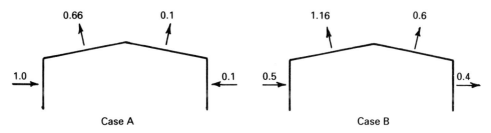

Fig. 5.25 Resultant pressure coefficients.

Case B: vertical load $= 0.318 \times 1.4 \times 6 = 2.671 \, \text{kN/m}$
wind load AB $= 0.5 \times 0.553 \times 1.4 \times 6 = 2.323 \, \text{kN/m}$
BC $= 1.16 \times 0.553 \times 1.4 \times 6 = 5.388 \, \text{kN/m}$
CD $= 0.6 \times 0.553 \times 1.4 \times 6 = 2.787 \, \text{kN/m}$
DE $= 0.4 \times 0.553 \times 1.4 \times 6 = 1.858 \, \text{kN/m}$

It is evident that Case A is likely to be the most critical and this will be analysed in full. The analysis for Case B would follow a similar course. A computer analysis for both cases is given in Appendix 5C.

5.12.2 Manual analysis of wind Case A

There are several manual methods available for this analysis. The work equation method provides an equally convenient alternative to the one used. Here we will adopt a procedure similar to that used for the initial design and illustrated in Fig. 5.3. The main difference is that drawing the free bending moment diagram is now a little more tedious. Figure 5.26 is a line diagram of the case to be considered where the uniformly distributed loads are those calculated above. The reactions arise from consideration of equilibrium as follows.

From horizontal equilibrium:

$$H = 17.92 - 1.79 - (29.73 - 4.50)\sin 6° = 13.49 \, \text{kN}$$

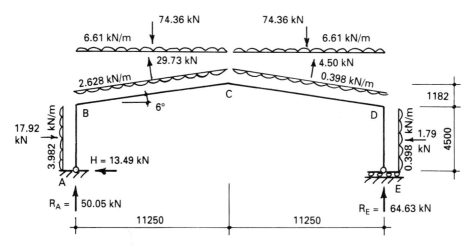

Fig. 5.26 Case to be considered when drawing the free bending moment diagram.

From moment equilibrium about A:

$$R_E = [(17.92 - 1.79)2.25 - 29.73\cos 6° \times 5.625 - 4.50\cos 6°$$

$$\times 16.875 - 29.73\sin 6° \times 5.091 + 4.50\sin 6°$$

$$\times 5.091 + 148.72 \times 11.25]/22.5$$

$$= 64.63\,\text{kN}$$

and from vertical equilibrium

$$R_A = 74.36 \times 2 - (29.73 + 4.50)\cos 6° - 64.63 = 50.05\,\text{kN}$$

The free bending moment diagram shown in Fig. 5.27 can now be drawn as follows:

With x measured from A to B

$$M_{AB} = -13.49x + \frac{3.982x^2}{2}$$

With x measured from B to C

$$M_{BC} = 17.919(2.25 + x\tan 6°) + 6.61\,\frac{x^2}{2} - \frac{2.628x^2}{2\cos^2 6°}$$

$$- R_A x - 13.49(4.5 + x\tan 6°)$$

$$= -20.385 - 49.585x + 1.9765x^2$$

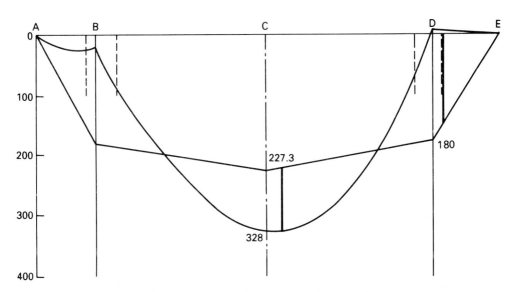

Fig. 5.27 Free and reactant bending moment diagrams for wind case A.

With x measured from E to D

$$M_{ED} = \frac{0.398x^2}{2}$$

and finally, with x measured from D to C

$$M_{DC} = 1.79(2.25 + x\tan 6°) + \frac{6.61x^2}{2} - \frac{0.398x^2}{2\cos^2 6°} - 64.63x$$

$$= 4.030 - 64.44x + 3.104x^2$$

It is a simple matter to check that the equations for M_{BC} and M_{DC} both give values of -328.1 kNm for $x = 11.25$ m at C.

Having drawn the free bending moment diagram to scale, as shown on Fig. 5.27, it is now only necessary to check that a statically admissible reactant line can be drawn that satisfies the yield criterion (no bending moment greater than the full plastic moment of the associated member). Only if wind should prove to be near-critical is it necessary to consider checking member stability under wind loading.

An arbitrary reactant line has been drawn on Fig. 5.27 for $H = 40$ kN. The only criterion for this choice of a value for H is that it gives a sensible looking bending moment diagram! The corresponding maximum bending moments scaled from the resulting diagram are:

For the stanchion: $M_{max} = 185$ kNm

For the rafter: $M_{max} = 105$ kNm

These are much smaller than the available capacities ($M_{ps} = 244$ kNm and $M_{pr} = 198$ kNm) so that no further check is necessary.

References

5.1 Steel Construction Institute. *Joints in steel construction – Moment connections*, Publication No. 207/95, 1995.
5.2 Morris L J & Plum D R *Structural steelwork design to BS 5950*. Longman Scientific and Technical, Harlow, 1988.

Appendix 5A   Approximate method for single-bay frames

This method considers the frame shown in Fig. 5A.1 and is taken from:

A D Weller. Portal frame design. Lecture 14 in Introduction to Steelwork Design to BS 5950: Part 1, Steel Construction Institute, reprint with Addendum, 1994.

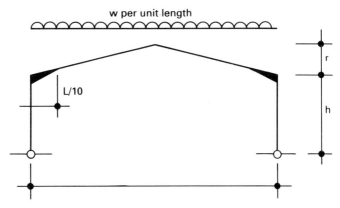

Fig. 5A.1 Symmetrical single-bay frame considered in the design method.

The graphs which follow have been produced to enable simple pinned based frames to be sized quickly. They are based on the following assumptions:

(1) Plastic hinges are formed at the bottom of the haunch in the leg and near the apex of the rafter, the exact position being determined by the frame geometry.

(2) The depth of the rafter is approximately span/55 and the depth of the haunch below the eaves intersection is 1.5 times the rafter depth.

(3) The haunch length is 10% of the span of the frame, a limit generally regarded as providing a balance between economy and stability.

(4) The moment in the rafter at the tip of the haunch is $0.87M_p$, so that it is assumed that the haunch remains elastic.

(5) The calculations assume that the calculated values of M_p are provided exactly by the available sections and there are no stability problems. Clearly, these conditions will not be precisely met and it is the responsibility of the designer to ensure that the chosen sections are checked for all aspects of behaviour.

The graphs cover the range of span/eaves height between 2 and 5 and rise/span of 0 to 0.2 (where 0 is a flat roof). Interpolation is permissible but extrapolation is not.

The three graphs give:

Fig. 5A.2   The horizontal thrust at the base of the frame as a proportion of the total factored load wL, where w is the load per unit length of rafter and L is the span of the frame.

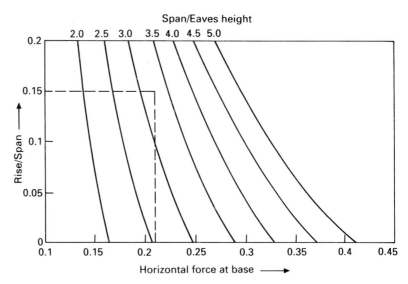

Fig. 5A.2 Rise/span versus horizontal base force for various values of span/eaves height.

Fig. 5A.3   The required moment capacity of the rafters as a proportion of wL^2.

Fig. 5A.4   The required moment capacity of the stanchions as a proportion of wL^2.

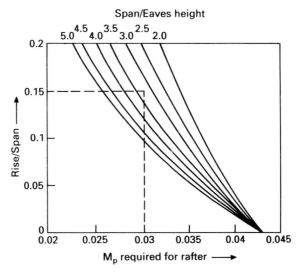

Fig. 5A.3 Rise/span versus required M_p of rafter for various values of span/eaves height.

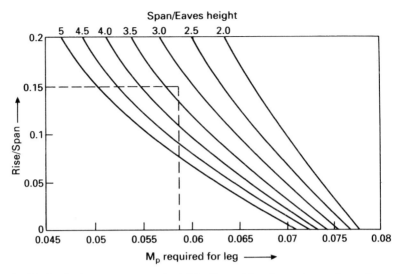

Fig. 5A.4 Rise/span versus required M_p of stanchion for various values of span/ eaves height.

The graphs are used as follows:

(1) Determine the ratio span/height to eaves (L/h).
(2) Determine the ratio rise/span (r/L).
(3) Calculate wL (total load) and wL^2.
(4) Determine the following values from the graphs:

Horizontal thrust at base	$= wL \times$ value from Fig. 5A.2.
M_p required for rafter	$= wL^2 \times$ value from Fig. 5A.3.
M_p required for stanchion	$= wL^2 \times$ value from Fig. 5A.4.

Appendix 5B   Elastic-plastic analysis of the designed frame

The analysis given in this Appendix was carried out using the core module of the market leading software package 'FASTRAK 5950'. The output is in the internal format of the program rather than the more graphically orientated information supplied by the post-processor available with the complete package.

The output provided by the program is in Courier font in order to preserve the tabulation and to distinguish it from the added notes. The first part of the output is a repeat of the input data for the purposes of identification and checking.

All of the data is in units of kN and mm.

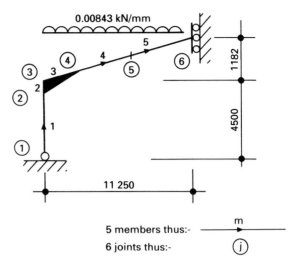

Fig. 5B.1 Line diagram for analysis.

Line diagram for analysis (Fig. 5B.1)

> *Note*:   As a consequence of symmetry, for the vertical load case it is only necessary to consider half of the frame.

Computer output

```
JOB NUMBER 1    PORTAL FRAME IN CHAPTER 5 (analysis case)

6  JOINTS      5  MEMBERS      0  YIELDING TIES

0  LOADED JOINTS      1  LOAD CASE      3  MEMBERS WITH UDL

                                    ALPHA-T      E
GLOBAL PARAMETERS IF RELEVANT       .000012      205.00
```

Joint data

The joint data is in the form of the degrees of freedom and the coordinates. The degrees of freedom are specified by the total number followed by an indication, in binary form and in the order x, y, θ, of the degrees of freedom which are active.

```
                                                     X           Y
JOINT NO 1 FREEDOM 1    1 COORDINATES             0.00     5682.00
JOINT NO 2 FREEDOM 3 111 COORDINATES              0.00     1782.00
JOINT NO 3 FREEDOM 3 111 COORDINATES              0.00     1182.00
JOINT NO 4 FREEDOM 3 111 COORDINATES           1350.00     1040.00
JOINT NO 5 FREEDOM 3 111 COORDINATES           6300.50      520.00
JOINT NO 6 FREEDOM 1   10 COORDINATES         11250.00        0.00
```

Member data

Members are specified by the joint numbers at 'end 1' and 'end 2' indicated by the arrows in the line diagram. The member data also specifies the 'critical sections' (C/S) where plastic hinges may form. 3 indicates a critical section that may move in order to locate a plastic hinge at the section of maximum bending moment. The length L and the inclination are calculated by the program as a check on the input data.

The member data also indicates a type number which identifies the relevant section properties from the member property table which follows.

MEMBER	JOINTS		C/S		TYPE	L	INCL
1	1	2	0	2	1	3900.00	−90.000
2	2	3	0	0	1	600.00	−90.000
3	3	4	0	0	1	1357.45	−6.005
4	4	5	2	3	2	4977.24	−5.997
5	5	6	0	2	2	4977.24	−5.997

4 CRITICAL SECTIONS

MEMBER PROPERTIES

TYPE	A	I	MP
1	5900.00	156700000.0	244000.00
2	4920.00	124100000.0	198000.00

Note: For illustrative purposes, the eaves haunch is modelled rather crudely by a single member with the same inertia as the stanchion. As no plastic hinge is allowed to form in this member, this is accurate for the strength calculation but rather approximate for the calculation of deflections. In practice, the haunch would be modelled more accurately by a small number of slices, each with an average inertia.

Load data

The program can accept a wide variety of forms of load data. This example only requires uniformly distributed vertical load which is applied to specified members as a load per unit length.

DISTRIBUTED LOADS

VERTICAL ON MEMBER	3	.8430E−02
VERTICAL ON MEMBER	4	.8430E−02
VERTICAL ON MEMBER	5	.8430E−02

Displacement output in the elastic-plastic phase

The user of the program has to specify which displacements are to be printed out during elastic-plastic analysis. Here we choose the horizontal eaves deflection and the vertical apex deflection specified by the joint number and degree of freedom.

```
2 LOCATIONS    3 1    6 2
```

Output from the analysis

Before the program enters the elastic-plastic analysis module, it first carries out an elastic analysis at unit load factor (i.e. at 1.0 times the required ultimate limit state loads). The deflections at the serviceability limit state can be determined from these pro rata.

```
DISPLACEMENTS AT JOINTS - LOAD CASE 1

                           X                Y              ROTATION
    JOINT    1       .00000E+00       .00000E+00        -.12157E-01
    JOINT    2      -.26766E+02       .30580E+00         .37252E-02
    JOINT    3      -.22990E+02       .35285E+00         .89880E-02
    JOINT    4      -.21007E+02       .20016E+02         .19350E-01
    JOINT    5      -.71233E+01       .15564E+03         .26468E-01
    JOINT    6       .00000E+00       .22670E+03         .00000E+00
```

```
MEMBER FORCES - LOAD CASE 1

                     END 1                                END 2
MEMBER     AXIAL       SHEAR      MOMENT       AXIAL        SHEAR      MOMENT
  1      .9484E+02   .6709E+02   .0000E+00  -.9484E+02  -.6709E+02   .2616E+06
  2      .9484E+02   .6709E+02  -.2616E+06  -.9484E+02  -.6709E+02   .3019E+06
  3      .7664E+02  -.8730E+02  -.3019E+06  -.7545E+02   .7598E+02   .1911E+06
  4      .7544E+02  -.7599E+02  -.1911E+06  -.7108E+02   .3449E+02  -.8388E+05
  5      .7108E+02  -.3449E+02   .8388E+05  -.6672E+02  -.7009E+01  -.1523E+06
```

Elastic-plastic phase of analysis

The program then traces the successive formations of plastic hinges as the global load factor is varied. The global load factor is a simple multiplier on the loads specified in the input data. For hinges which relocate at the section of maximum bending moment, ALPHA gives the hinge position as a proportion of the total length of the member measured from the left-hand end.

SET HINGE 1   LOAD FACTOR   .932575E+00

CURRENT BENDING MOMENTS AT 4 CRITICAL SECTIONS

	MP	M	UNIT MOMENT	AXIAL	SHEAR FORCE
1	.244000E+06	−.244000E+06	−.261641E+06	.88443E+02	.62564E+02
2	.198000E+06	−.178188E+06	−.191071E+06	.70353E+02	−.70868E+02
3	.198000E+06	.782220E+05	.838774E+05	.66287E+02	−.32166E+02
4	.198000E+06	.142003E+06	.152270E+06	.62222E+02	.65364E+01

DISPLACEMENTS AT FORMATION OF FIRST HINGE
     −.2144E+02   .2114E+03

TRAVELLING HINGE,   ALPHA= .91639   MP=198000.0000

SET HINGE 3   LOAD FACTOR   .103294E+01

CURRENT LIMITING BENDING MOMENTS AT THE CRITICAL SECTIONS

	MP	M	AXIAL	SHEAR FORCE
1	.244000E+06	−.244000E+06	.979612E+02	.625641E+02
2	.198000E+06	−.166110E+06	.712281E+02	−.791977E+02
3	.198000E+06	.198000E+06	.629747E+02	.631761E+00
4	.198000E+06	.195543E+06	.622217E+02	.653642E+01

DISPLACEMENTS FOR EACH LOAD SYSTEM AS SPECIFIED
     −.3313E+02   .3230E+03

CURRENT ROTATIONS OF THE PLASTIC HINGES
     .2124E−01   .0000E+00

MECHANISM CRITERION SATISFIED

THE FINAL MECHANISM IS CONFIRMED AS VALID
FAILURE LOAD=1.0329

Plastic hinges at collapse (Fig. 5B.2 showing critical section numbers)

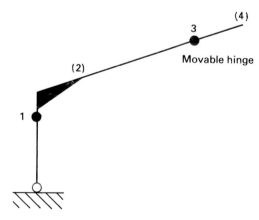

Fig. 5B.2 Plastic hinges at collapse.

Appendix 5C   Elastic-plastic analysis for the wind load cases

The general details of the analysis used are given in Appendix 5B. Additional details are given here only when they are additional to those given previously.

Line diagram for analysis

Here it is necessary to analyse the whole frame, as shown in Fig. 5C.1, because the wind load cases are not symmetrical.

Computer output

```
JOB NUMBER  2     PORTAL FRAME IN CHAPTER 5 (wind load cases)

11  JOINTS      10  MEMBERS       0  YIELDING TIES

0  LOADED JOINTS      2  LOAD CASES      16  MEMBERS WITH UDL

                                      ALPHA-T           E
GLOBAL PARAMETERS IF RELEVANT       0.000012        205.00

JOINT DATA
                                                   X            Y
JOINT NO  1 FREEDOM 1    1 COORDINATES         0.00      5682.00
JOINT NO  2 FREEDOM 3 111 COORDINATES          0.00      1782.00
JOINT NO  3 FREEDOM 3 111 COORDINATES          0.00      1182.00
JOINT NO  4 FREEDOM 3 111 COORDINATES       1350.00      1040.00
JOINT NO  5 FREEDOM 3 111 COORDINATES       6300.00       520.00
JOINT NO  6 FREEDOM 3 111 COORDINATES      11250.00         0.00
JOINT NO  7 FREEDOM 3 111 COORDINATES      16200.00       520.00
JOINT NO  8 FREEDOM 3 111 COORDINATES      21150.00      1040.00
JOINT NO  9 FREEDOM 3 111 COORDINATES      22500.00      1182.00
JOINT NO 10 FREEDOM 3 111 COORDINATES      22500.00      1782.00
JOINT NO 11 FREEDOM 1    1 COORDINATES      22500.00      5682.00
```

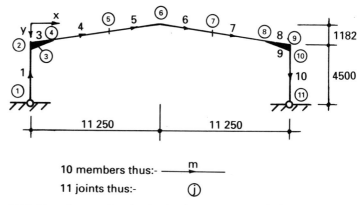

Fig. 5C.1 Line diagram for elastic-plastic computer analysis.

MEMBER DATA

MEMBER	JOINTS		C/S		TYPE	L	INCL
1	1	2	0	2	1	3900.00	-90.000
2	2	3	0	0	1	600.00	-90.000
3	3	4	0	0	1	1357.45	-6.005
4	4	5	2	3	2	4977.24	-5.997
5	5	6	0	2	2	4977.24	-5.997
6	6	7	0	3	2	4977.24	5.997
7	7	8	0	2	2	4977.24	5.997
8	8	9	0	0	1	1357.45	6.005
9	9	10	0	0	1	600.00	90.000
10	10	11	2	0	1	3900.00	90.000

7 CRITICAL SECTIONS

MEMBER PROPERTIES

TYPE	A	I	MP
1	5900.00	156700000.0	244000.00
2	4920.00	124100000.0	198000.00

Load data

For this analysis, the load data includes uniformly distributed loads which are either vertical or normal to the member. Both are in units of kN/mm. The loads for the two load cases are expressed columnwise.

		LOAD CASE 1	LOAD CASE 2
VERTICAL ON MEMBER	3	.6610E-02	.1908E-02
VERTICAL ON MEMBER	4	.6610E-02	.1908E-02
VERTICAL ON MEMBER	5	.6610E-02	.1908E-02
VERTICAL ON MEMBER	6	.6610E-02	.1908E-02
VERTICAL ON MEMBER	7	.6610E-02	.1908E-02
VERTICAL ON MEMBER	8	.6610E-02	.1908E-02
NORMAL ON MEMBER	1	.3982E-02	.2323E-02
NORMAL ON MEMBER	2	.3982E-02	.2323E-02
NORMAL ON MEMBER	3	-.2628E-02	-.5388E-02
NORMAL ON MEMBER	4	-.2628E-02	-.5388E-02
NORMAL ON MEMBER	5	-.2628E-02	-.5388E-02
NORMAL ON MEMBER	6	-.3980E-03	-.2787E-02
NORMAL ON MEMBER	7	-.3980E-03	-.2787E-02
NORMAL ON MEMBER	8	-.3980E-03	-.2787E-02
NORMAL ON MEMBER	9	.3980E-03	-.1858E-02
NORMAL ON MEMBER	10	.3980E-03	-.1858E-02

DISPLACEMENTS OUTPUT IN ELASTIC-PLASTIC PHASE
3 LOCATIONS   3 1   6 2   9 1

Output from the analysis

```
           DISPLACEMENTS AT JOINTS - LOAD CASE 1

                          X              Y           ROTATION
        JOINT     1    .00000E+00     .00000E+00    -.65989E-02
        JOINT     2   -.15771E+02     .16140E+00     .13723E-02
        JOINT     3   -.14106E+02     .18623E+00     .42647E-02
        JOINT     4   -.13110E+02     .10194E+02     .10149E-01
        JOINT     5   -.53510E+01     .86360E+02     .15764E-01
        JOINT     6   -.58111E+00     .13391E+03     .17446E-02
        JOINT     7    .30278E+01     .97463E+02    -.15563E-01
        JOINT     8    .11641E+02     .13222E+02    -.12607E-01
        JOINT     9    .12950E+02     .24046E+00    -.60857E-02
        JOINT    10    .15570E+02     .20839E+00    -.27258E-02
        JOINT    11    .00000E+00     .00000E+00     .73360E-02


           DISPLACEMENTS AT JOINTS -- LOAD CASE 2

                          X              Y           ROTATION
        JOINT     1    .00000E+00     .00000E+00     .35658E-02
        JOINT     2    .67248E+01    -.10647E+00    -.17799E-02
        JOINT     3    .51783E+01    -.12285E+00    -.34048E-02
        JOINT     4    .44814E+01    -.69485E+01    -.64279E-02
        JOINT     5    .43328E+00    -.46394E+02    -.66179E-02
        JOINT     6   -.67448E+00    -.57896E+02     .20346E-02
        JOINT     7   -.31361E+01    -.33445E+02     .68520E-02
        JOINT     8   -.61885E+01    -.34176E+01     .35612E-02
        JOINT     9   -.65193E+01    -.59597E-01     .12819E-02
        JOINT    10   -.69545E+01    -.51649E-01     .20166E-03
        JOINT    11    .00000E+00     .00000E+00    -.27042E-02
```

```
MEMBER FORCES - LOAD CASE 1

                     END 1                              END 2
MEMBER    AXIAL      SHEAR       MOMENT       AXIAL      SHEAR       MOMENT
  1     .5005E+02  .2849E+02  -.3183E-10  -.5005E+02  -.4402E+02   .1414E+06
  2     .5005E+02  .4402E+02  -.1414E+06  -.5005E+02  -.4641E+02   .1685E+06
  3     .5139E+02 -.4492E+02  -.1685E+06  -.5046E+02   .3962E+02   .1112E+06
  4     .5046E+02 -.3962E+02  -.1112E+06  -.4704E+02   .2016E+02  -.3762E+05
  5     .4704E+02 -.2016E+02   .3762E+05  -.4362E+02   .7023E+00  -.8955E+05
  6     .4252E+02 -.9751E+01   .8955E+05  -.4594E+02  -.2081E+02  -.6203E+05
  7     .4594E+02  .2081E+02   .6203E+05  -.4936E+02  -.5137E+02   .1176E+06
  8     .4936E+02  .5136E+02  -.1176E+06  -.5030E+02  -.5970E+02   .1930E+06
  9     .6463E+02 -.4378E+02  -.1930E+06  -.6463E+02   .4354E+02   .1668E+06
 10     .6463E+02 -.4354E+02  -.1668E+06  -.6463E+02   .4199E+02   .4434E-10
```

```
MEMBER FORCES - LOAD CASE 2

                     END 1                              END 2
MEMBER    AXIAL      SHEAR       MOMENT       AXIAL      SHEAR       MOMENT
  1    -.3302E+02 -.2560E+02  -.3638E-11   .3302E+02   .1654E+02  -.8217E+05
  2    -.3302E+02 -.1654E+02   .8217E+05   .3302E+02   .1515E+02  -.9168E+05
  3    -.1852E+02  .3125E+02   .9168E+05   .1879E+02  -.2650E+02  -.5248E+05
  4    -.1878E+02  .2651E+02   .5248E+05   .1977E+02  -.9081E+01   .3608E+05
  5    -.1977E+02  .9081E+01  -.3608E+05   .2076E+02   8344E+01    .3792E+05
```

```
  6   −.2204E+02  −.3848E+01  −.3792E+05   .2105E+02   .8327E+01   .7617E+04
  7   −.2105E+02  −.8327E+01  −.7617E+04   .2006E+02   .1281E+02  −.4497E+05
  8   −.2007E+02  −.1280E+02   .4497E+05   .1980E+02   .1402E+02  −.6318E+05
  9   −.1602E+02   .1822E+02   .6318E+05   .1602E+02  −.1711E+02  −.5258E+05
 10   −.1602E+02   .1711E+02   .5258E+05   .1602E+02  −.9860E+01  −.1910E−10
```

ENTER ELASTIC-PLASTIC PHASE OF THE ANALYSIS

LOAD CASE   1

SET HINGE   7   LOAD FACTOR   .146312E+01

CURRENT BENDING MOMENTS AT 7 CRITICAL SECTIONS

```
          MP              M           UNIT MOMENT        AXIAL       SHEAR FORCE
 1   .244000E+06   −.206897E+06   −.141409E+06     .73233E+02     .64412E+02
 2   .198000E+06   −.162641E+06   −.111161E+06     .73822E+02    −.57973E+02
 3   .198000E+06    .550454E+05    .376220E+05     .68820E+02    −.29500E+02
 4   .198000E+06    .131017E+06    .895465E+05     .63819E+02    −.10275E+01
 5   .198000E+06    .907564E+05    .620295E+05     .67214E+02     .30445E+02
 6   .198000E+06   −.172047E+06   −.117589E+06     .72215E+02     .75157E+02
 7   .244000E+06   −.244000E+06   −.166767E+06     .94560E+02    −.63701E+02
```

DISPLACEMENTS AT FORMATION OF FIRST HINGE
```
      −.2064E+02        .1959E+03        .1895E+02
```

TRAVELLING HINGE ALPHA=0.15882   MP=198000.0

SET HINGE   5   LOAD FACTOR   .163014E+01

CURRENT LIMITING BENDING MOMENTS

```
          MP              M             AXIAL       SHEAR FORCE
 1   .244000E+06   −.202661E+06    .815936E+02     .646223E+02
 2   .198000E+06   −.148053E+06    .751460E+02    −.653371E+02
 3   .198000E+06    .981988E+05    .695736E+02    −.336140E+02
 4   .198000E+06    .186557E+06    .640011E+02    −.189099E+01
 5   .198000E+06    .198000E+06    .654868E+02     .118270E+02
 6   .198000E+06   −.158532E+06    .733559E+02     .844832E+02
 7   .244000E+06   −.244000E+06    .105355E+03    −.638308E+02
```

DISPLACEMENTS AS SPECIFIED
```
      .5489E+02        .3125E+03        .1189E+03
```

CURRENT PLASTIC HINGE ROTATIONS
```
      .0000E+00        .4506E−01
```

MECHANISM CRITERION SATISFIED

THE FINAL MECHANISM IS CONFIRMED AS VALID

FAILURE LOAD=1.6301

NEW LOAD CASE 2

SET HINGE   1   LOAD FACTOR   .296933E+01

CURRENT BENDING MOMENTS

	MP	M	UNIT MOMENT	AXIAL	SHEAR FORCE
1	.244000E+06	.244000E+06	.821733E+05	−.98049E+02	−.49113E+02
2	.198000E+06	.155824E+06	.524779E+05	−.55774E+02	.78703E+02
3	.198000E+06	−.107139E+06	−.360817E+05	−.58704E+02	.26964E+02
4	.198000E+06	−.112584E+06	−.379157E+05	−.61634E+02	−.24775E+02
5	.198000E+06	−.226170E+05	−.761687E+04	−.62507E+02	−.24725E+02
6	.198000E+06	.133541E+06	.449732E+05	−.59577E+02	−.38024E+02
7	.244000E+06	.156132E+06	.525817E+05	−.47563E+02	.50793E+02

DISPLACEMENTS AT FORMATION OF FIRST HINGE
      .1538E+02   −.1719E+03   −.1936E+02

TRAVELLING HINGE ALPHA=0.76120 MP=−198000.0

SET HINGE 3   LOAD FACTOR .331361E+01

CURRENT LIMITING BENDING MOMENTS

	MP	M	AXIAL	SHEAR FORCE
1	.244000E+06	.244000E+06	−.109417E+03	−.475540E+02
2	.198000E+06	.140219E+06	−.550268E+02	.85855E+02
3	.198000E+06	−.198000E+06	−.600045E+02	.685635E+00
4	.198000E+06	−.166854E+06	−.615661E+02	−.268900E+02
5	.198000E+06	−.626839E+05	−.625404E+02	−.283496E+02
6	.198000E+06	.115351E+06	−.592708E+02	−.431901E+02
7	.244000E+06	.145946E+06	−.530782E+02	.494281E+02

DISPLACEMENTS AS SPECIFIED
      .1165E+03   −.2875E+03   .5749E+02

CURRENT PLASTIC ROTATIONS
      −.4576E−01   .0000E+00

THE FINAL MECHANISM IS CONFIRMED AS VALID

FAILURE LOAD=3.314

Chapter 6
Plastic Design of Multi-storey Buildings

6.1 General

Clause 5.1.3, BS 5950, Part 1 makes a fundamental distinction between *sway* and *non-sway* multi-storey frames. This is potentially confusing because this distinction is nothing to do with the way the frame is braced, it is rather a matter of the *sway stiffness* of the frame. A multi-storey frame may be classed as non-sway, whether or not it is braced, if its sway stiffness is sufficient for second-order effects to be neglected. A simple rule is given for making this distinction:

'A rigid jointed multi-storey frame may be considered as a non-sway frame if in every individual storey the deflection δ in storey height h, due to the notional horizontal loading given in 5.1.2.3 (0.5% of the factored dead plus vertical imposed load at each storey applied horizontally) satisfies the following criteria,

(1) For clad frames where the stiffening effect of the cladding is not taken into account in the deflection calculations:

$$\delta \leq \frac{h}{2000}$$

(2) For unclad frames or clad frames where the stiffening effect of the cladding is taken into account in the deflection calculations:

$$\delta \leq \frac{h}{4000}$$

A rigid jointed multi-storey frame which does not comply with the above criteria should be classed as a sway frame even if it is also braced.'

It may be noted that this rule effectively requires that the elastic critical load factor, calculated according to the deflection method given in

BS 5950, Appendix F, should be greater than 10 for clad frames or 20 for unclad frames. The background to this requirement is discussed in Section 3.2.1 of Chapter 3.

The distinction between sway and non-sway frames is even more important in plastic design than in elastic design because, as successive plastic hinges form, the rotational stiffness at the plastic hinge position vanishes and therefore the frame displacements are increased and second-order effects enhanced. It follows that, in general, it is desirable to ensure that plastically designed multi-storey buildings are non-sway in the terms of the above clause. This can be achieved by ensuring that a frame which is free to sway is sufficiently stiff for it to be classed as non-sway. However, the required stiffness is more likely to be achieved by the provision of a triangulated vertical bracing system or by relatively rigid reinforced concrete shear walls or core walls around lift shafts and stairwells acting in conjunction with concrete floors acting as stiff horizontal diaphragms. Each frame must also be effectively braced against sidesway out of its plane.

Beam-to-stanchion connections at plastic hinge positions must be designed to have adequate moment rotation characteristics or be designed to be over-strong so that the plastic hinges occur only in the basic beam cross-section.

Designers of multi-storey buildings should also be aware of BS 5950, clause 2.4.5.3 'Additional requirements for certain multi-storey buildings'. This clause draws attention to the fact that local or national regulations may stipulate that tall multi-storey buildings should be designed to localise accidental damage and gives guidance on how this may be achieved.

Pattern loading in multi-storey buildings may pose particular design problems. For the purposes of this chapter, a pragmatic design approach is used whereby, for load combinations involving horizontal loads, pattern loading of vertical loads need not be considered. Without this realistic rule, it is necessary to have specialist computer programs for the plastic design of multi-storey frames.

6.2   Non-sway frames

If a multi-storey frame can be classified as non-sway, BS 5950: Part 1 contains no design requirements other than that the buckling resistance of the members should be checked by reference to clause 4.8.3.3. This clause is only appropriate when it can be shown that the member being checked does not contain a plastic hinge that requires plastic rotational capacity at frame loads below the required ultimate limit state loading. It is reasonable to argue that a plastic hinge history of the frame for each loading combination to determine the forming and unforming of any transient hinges is not required because such hinges are unlikely to require full plastic hinge rotational capacity and therefore 4.8.3.3 is still appropriate. However, in order to justify the use of clause 4.8.3.3, a permanent plastic hinge should not form in a stanchion at much below the required ultimate limit state load. Where plastic hinges do form in the stanchions at relatively low load levels, then they should be checked in accordance with more suitable criteria such as Reference 3.15. Reference 3.16, which is discussed in Section 3.3.3 of this book, gives a suitable simplified method for Universal Beam sections.

Sub-frames are permitted in the elastic design of rigid frames under vertical load combinations as described in BS 5950, clause 5.6.4. There is no mention of sub-frames in the clauses concerned with the plastic design of multi-storey frames. However, there may be the need to analyse pattern loading combinations where member plastic hinge mechanisms occur adjacent to elastic members. Furthermore, a non-sway condition is one of the design checks required for sway frames and it is considered to be reasonable to use sub-frames in the analysis of non-sway frames subject to vertical load combinations. Sub-frame layouts, different from those used in elastic analysis, are recommended and have been included in the design example 6.1.

6.3   Sway frames

The plastic design of sway frames is permitted by BS 5950 provided proper allowance is made for frame instability effects. This may be done by carrying out a full elastic-plastic sway analysis or by using the simplified method in clause 5.7.3.3.

Plastic hinges in stanchions are only acceptable if full elastic-plastic sway analyses are carried out. These analyses require the use of specialist computer programs which should produce a full plastic hinge history including transient hinges, moment rotational requirements, particularly at connection hinges, and include second-order effects by the use of stability functions or other appropriate theories. Specialist programs should be approved by expert third parties because the results are not readily verifiable by practising engineers.

In the plastic design of sway frames the use of sub-frames should be limited to frames of four storeys or less. This means in practice that computer programs are recommended for frames with five or more storeys. However, these programs need not include second-order effects if they are used with the simplified method. This chapter only deals with the simplified method.

6.3.1 Design procedure for sway frames using the simplified method

The following steps are required in the plastic design of a sway frame. These steps are then illustrated in the design example which follows.

(1) Carry out a preliminary design of the frame in the absence of second-order effects taking into account all (reasonable) combinations of loading. Assume that the effective length of all stanchions in the plane of the frame is $1.0 \times L$.

   When considering vertical loading in the absence of wind load, then the notional horizontal loading should be applied.

(2) The simplified check for frame stability given in BS 5950, clause 5.7.3.3 should next be satisfied, thus:

   (a) The plastic hinge mechanism should be a sway mode with plastic hinges assumed in all beams and at the base of each stanchion. No other plastic hinges are allowed in the stanchions.

   (b) The individual beams should be checked to ensure that there are no local beam mechanisms at the design loads.

   (c) The lower lengths of the stanchions should be designed to remain elastic under the theoretical hinge moments assumed in (a).

   (d) Under all (reasonable) combinations of unfactored loading, it should be possible by means of moment redistribution to ensure that all members remain elastic.

   (e) Second-order effects are accounted for by increasing the minimum overall rigid-plastic load factor to satisfy (a) and (c) depending on the elastic critical load factor as determined using Appendix F of BS 5950.

(3) Even though, for any reason, the strength of the frame may be increased above the minimum required for strength design, the stability of the stanchions may be checked using the factored loads (with bending moments and shear forces enhanced as necessary in accordance with 2(e)) rather than the actual collapse loads of the frame.

(4) The stanchions should be checked under vertical pattern loading and, for this purpose, the frame may be assumed to be non-sway. The in-plane stanchion effective lengths can be derived from

Appendix E of BS 5950, but usually the out-of-plane slenderness will be critical if restraint is provided only at the floor levels.

6.4   Example 6.1. Design of a sway frame

6.4.1   General

A four-storey rigid frame is shown in Fig. 6.1. It has nominally pinned bases and a pin jointed roof structure, hence the roof is not part of this section of the design. The frames are at 6 m centres.

6.4.2   Loading

6.4.2.1   General
The unfactored loadings assumed are shown in Table 6.1.

The reduction in imposed floor loading is in accordance with Table 3 of BS 6399: Part 1: 1984. A 6% reduction is permitted on each beam supporting an area of 60 m^2.

The unfactored weight of the side stanchions and wall construction has been taken as 12.5 kN per metre length of stanchion.

6.4.2.2   Stanchion loading reductions
Table 2 of BS 6399: Part 1 gives reductions in imposed loading of 30, 20, 10 and 0% on the 1st, 2nd, 3rd and 4th storey stanchions respectively. This is *not* in addition to the 6% reduction on the floor beams indicated above.

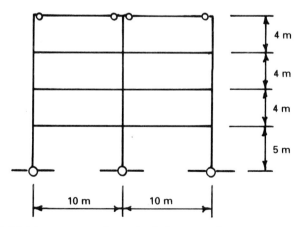

Fig. 6.1 Multi-storey sway frame in design example.

Table 6.1 Unfactored design loads for four-storey frame.

		kN/m²	kN/m	kN/m
Roof	Imposed	0.75	4.5	
	Dead	3.0	18.0	
		3.75	22.5	
Floors	Imposed	5.0 ⎱	36.0 reduced to:	33.84
Partitions	Imposed	1.0 ⎰		
	Dead	4.0	24.0	24.0
		10.0	60.0	57.84

6.4.2.3 Wind loading

The wind loading has been derived from CP3: Chapter V: Part 2: 1972.

Basic wind speed assumed $= 44\,\text{m/s}$.

Topography factor	$S_1 = 1.0$	
Ground roughness factor	S_2	based on Category 3 Class C with the factor calculated for each storey as given below.
Statistical factor	$S_3 = 1.0$	

The storey panel loads may be calculated as shown in Table 6.2.
   The wind loads at roof and floor levels are therefore:

Roof	$24.06 \times$ say 0.6	$= 14.44\,\text{kN}$
3rd floor	$(15.80 + 24.06 \times 0.8)/2$	$= 17.52\,\text{kN}$
2nd floor	$(12.90 + 15.80)/2$	$= 14.35\,\text{kN}$
1st floor	$(12.84 + 12.90)/2$	$= 12.87\,\text{kN}$

The wind loads on the roof have not been included in this design example. The loads on the frame are based on a force coefficient including windward and leeward walls of $(0.7 + 0.3)q = 1.0q$.

Table 6.2 Calculation of storey panel loads.

Storey	Height (m)	S_2	Design wind speed (m/s)	q (kN/m²)	Storey panel loads (kN)
1	5	0.600	26.40	0.428	12.84
2	9	0.672	29.57	0.537	12.90
3	13	0.744	32.74	0.658	15.80
4 say	18	0.822	36.17	0.802	24.06

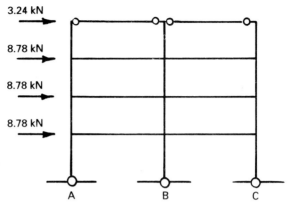

Fig. 6.2 Notional loads on frame for load case 1.

Table 6.3 Axial loads in stanchions in kN for load case 1.

	A	B	C
4th storey	162	324	162
13/20	−1		+1
	70	12	70
	231	336	233
3rd storey	231	336	233
	456	912	456
48/20	−2		+2
	−29	−58	−29
	70	12	70
	726	1202	732
2nd storey	726 + 29 = 755	1202 + 58 = 1260	732 + 29 = 761
	456	912	456
84/20	−4		+4
	−115	−230	−115
	70	12	70
	1162	1954	1176
1st storey	1162 + 115 = 1277	1954 + 230 = 2184	1176 + 115 = 1291
	456	912	456
147/20	−7		+7
	−259	−518	−259
	88	15	88
	1555	2593	1583

6.4.2.4   Notional horizontal loads

For load combinations with vertical loads, in the absence of wind loads, and in the derivation of the elastic critical load factor using clauses 5.1.2.3, 5.6.3 and Appendix F, the notional horizontal load at each floor and roof level is 0.5% of the factored vertical loads applied at that level as shown in Fig. 6.2.

For the roof $N_r = 0.005\,(1.4 \times 18.0 + 1.6 \times 4.5)\,20 = 3.24\,\text{kN}$

For each floor $N_f = 0.005\,(1.4 \times 24.0 + 1.6 \times 33.84)\,20 = 8.78\,\text{kN}$

Loading summaries, including stanchion axial loads, are given in Tables 6.3 and 6.4 for the two ultimate limit state load cases:

Case 1 $1.4 \times \text{Dead} + 1.6 \times \text{Imposed (all floors loaded)} + \text{Notional loads}$
Case 2 $1.2 \times \text{Dead} + 1.2 \times \text{Imposed (all floors loaded)} + 1.2 \times \text{Wind}$.

Loading summary for Case 1, $1.4 \times Dead + 1.6 \times Imposed + Notional$
horizontal loads (Fig. 6.2 and Table 6.3)

Notional loads per storey	Notional load shear moments per storey
3.24 kN	13 kNm
3.24	
8.78	
12.02	48 kNm
12.02	
8.78	
20.80	84 kNm
20.80	
8.78	
29.58	147 kNm

For the axial loads in the stanchions, the gross imposed load is used and the BS 6399 reductions are made later.

Roof $(1.4 \times 18 + 1.6 \times 4.5)\,10 = 324\,\text{kN}$
Floor $(1.4 \times 24 + 1.6 \times 36)\,10 = 912\,\text{kN}$

The imposed component of floor load to be reduced is $1.6 \times 36 \times 10 = 576\,\text{kN}$.

Table 6.4 Axial loads in stanchions in kN for load case 2.

	A	B	C
4th storey	135	270	135
69/20	−3		+3
	60	10	60
	$\overline{192}$	$\overline{280}$	$\overline{198}$
3rd storey	192	280	198
	360	720	360
154/20	−8		+8
	−21	−43	−21
	60	10	60
	$\overline{583}$	$\overline{967}$	$\overline{605}$
2nd storey	583 + 21 = 604	967 + 43 = 1010	605 + 21 = 626
	360	720	360
222/20	−11		+11
	−86	−172	−86
	60	10	60
	$\overline{927}$	$\overline{1568}$	$\overline{971}$
1st storey	927 + 86 = 1013	1568 + 172 = 1740	971 + 86 = 1057
	360	720	360
355/20	−18		+18
	−194	−388	−194
	72	15	72
	$\overline{1233}$	$\overline{2087}$	$\overline{1313}$

3rd storey stanchion reduction = 10% of one floor = 58 kN
2nd storey = 20% of two floors = 230 kN
1st storey = 30% of three floors = 518 kN

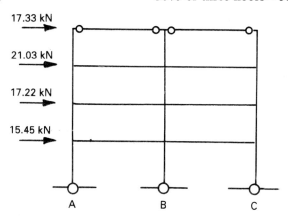

Fig. 6.3 Wind loads on frame for load case 2.

The above stanchion reduction is for the centre stanchion, hence take 50% of this for each side stanchion.

The resulting design axial loads in the stanchions, calculated as $1.4 \times$ Dead $+ 1.6 \times$ Imposed $+$ Notional horizontal loads, are given in Table 6.3.

Loading summary for Case 2, $1.2 \times$ Dead $+ 1.2 \times$ Imposed $+ 1.2 \times$ Wind load (Fig. 6.3 and Table 6.4)

The design wind loads for this load case are shown in Fig. 6.3.

Wind loads per storey	Wind load shear moments per storey
17.33 kN	69 kNm
17.33 21.03 ———— 38.36	154 kNm
38.36 17.22 ———— 55.58	222 kNm
55.58 15.45 ———— 71.03	355 kNm

For the axial loads in the stanchions, the gross imposed load is used and the BS 6399 reductions are made later.

Roof $(1.2 \times 18 + 1.2 \times 4.5)\, 10 = 270\, \text{kN}$
Floor $(1.2 \times 24 + 1.2 \times 36)\, 10 = 720\, \text{kN}$

The imposed component of the floor load to be reduced is $1.2 \times 36 \times 10 = 432$ kN.

3rd storey stanchion reduction $= 10\%$ of one floor $= 43\, \text{kN}$
2nd storey $= 20\%$ of two floors $= 172\, \text{kN}$
1st storey $= 30\%$ of three floors $= 388\, \text{kN}$

The above reduction is for the centre stanchion, hence take 50% of this for each side stanchion.

The resulting design axial loads in the stanchions, calculated as $1.2 \times$ Dead $+ 1.2 \times$ Imposed $+ 1.2 \times$ wind loads, are given in Table 6.4.

6.4.3   Preliminary member sizing

6.4.3.1   General

A sway mode plastic hinge mechanism can be used for preliminary member sizing with the requirement that all plastic hinges occur in the beams and, if nominally fixed bases are used, at the base of each stanchion.

In Fig. 6.4, the assumed loading and mechanism are shown for load case 1, $1.4 \times \text{Dead} + 1.6 \times \text{Imposed} + \text{Notional}$ loads.

A preliminary value for the required plastic moment of resistance of the beams can be found by solving the work equation for the sidesway mechanism shown.

$$8.78(5 + 9 + 13)\theta + 3.24 \times 17\theta + 6 \times 878 \times \frac{x\theta}{2} = 12M_{pb}(\theta + \alpha)$$

but $(\theta + \alpha) = \left(\dfrac{L}{L - x}\right)\theta$ with $L = 10\,\text{m}$

so that

$$M_{pb} = \frac{(292.14 + 2634x)(10 - x)}{120}$$

The critical plastic hinge position is given by

$$\frac{dM_{pb}}{dx} = 26\,047.9 - 5268x = 0$$

i.e. $x = 4.945\,\text{m}$ and $M_{pb} = 561.0\,\text{kNm}$

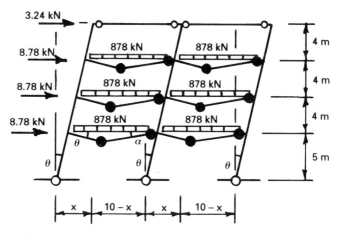

Fig. 6.4 Sidesway mechanism for member sizing, load case 1.

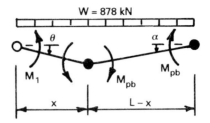

Fig. 6.5 Virtual mechanism to calculate M_1.

The bending moment M_1 at the left-hand end of each beam does not have a plastic hinge moment but it can be calculated by applying the virtual work equation to the mechanism shown in Fig. 6.5.

$$W \frac{x\theta}{2} = M_1\theta + M_{\text{pb}}(\theta + 2\alpha)$$

i.e. $M_1 = \dfrac{Wx}{2} - M_{\text{pb}}\left(\dfrac{L+x}{L-x}\right)$

$$= \frac{878 \times 4.945}{2} - 561.0 \times \frac{14.945}{5.055} = 512.3\,\text{kNm}$$

The value of M_1 is less than $M_{\text{pb}} = 561\,\text{kNm}$ under this, the load case which subjects the beams to the greatest intensity of vertical load. This ensures that the structure is not subject to failure in a local beam mechanism.

Similarly, for load case 2, namely $1.2 \times \text{Dead} + 1.2 \times \text{Imposed} + 1.2 \times \text{Wind}$, the mechanism for member sizing is shown in Fig. 6.6

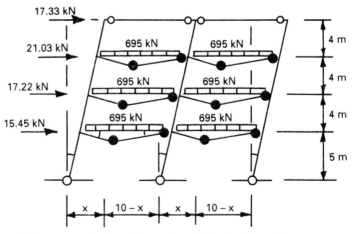

Fig. 6.6 Sidesway mechanism for member sizing, load case 2.

which gives the work equation

$$\left(15.45 \times 5 + 17.22 \times 9 + 21.03 \times 13 + 17.33 \times 17 + 6 \times 695 \times \frac{x}{2} \right) \theta$$

$$= 12 M_{pb}(\theta + \alpha)$$

so that

$$M_{pb} = \frac{(800.2 + 2085x)(L - x)}{12L} \quad \text{with} \quad L = 10 \text{ m}$$

and hence, $x = 4.808$ m, $M_{pb} = 468.4$ kNm and $M_1 = 335.0$ kNm.

6.4.3.2   Beam sizing

The required preliminary plastic moment of resistance for the beams is therefore 561.0 kNm to which must be added some allowance for second-order effects which can be as much as an additional 15%.

Adopt as beams $533 \times 210 \times 92$ UB in design Grade 43 steel which has a plastic moment of 651 kNm. This gives an additional load factor against collapse of $651/561 = 1.16$ according to first-order (rigid plastic) theory. It is shown later in Section 6.4.4 that this is sufficient to take account of second-order effects.

6.4.3.3   Stanchion sizing

When choosing appropriate section sizes for the stanchions, a decision has to be made as to whether or not to try to use smaller stanchion section sizes for the top two storeys. In making this decision, the points to consider are:

(1) The stanchions in rigid frames attract large bending moments from the floor beams irrespective of the floor level in the frame.

(2) Plastic hinges forming in the upper stanchions early in the hinge history may well allow a local mechanism to form at a load level below the required ultimate limit state loading.

(3) Internal stanchions will attract bending moments from pattern loading.

(4) Splices in the stanchions of rigid frames will have to have large bending moment and axial load capacities and will require non-slip connections. This almost certainly means the use of friction grip bolts together with all of the additional costs that such connections incur.

(5) Extra design office and drawing office time and smaller tonnages of more different steel sections will also add to the costs.

(6) Beam-to-stanchion connections are moment resisting connections and therefore require reasonably large stanchion flange thicknesses in order to resist the high tensile loads from the bolts.

(7) Splices increase the footprint of the stanchions.

Conversely, however, 17 m long stanchions will require careful consideration of the straightness and section tolerances and an

Table 6.5 Plastic moment distribution of stanchion moments for load case 1.

Shear moments per storey				
	246	−14	−244	Final moments
	250	−10	−240	
12 kNm	−4	−4	−4	
3rd floor	A3	B3	C3	
	−8 (−512)	(49) −8	(561) −8	
	12	−26	−248	
	262		−62	
	266	−34	−318	Final moments
48 kNm				
	292	−8	−246	Final moments
	300	0	−300	
			+62	
	−8	−8	−8	
2nd floor	A2	B2	C2	
	−14 (−512)	(49) −14	(561) −14	
	22	−26	−208	
	212		−94	
	220	−40	−316	Final moments
84 kNm				
	286	−14	−220	Final moments
	300	0	−300	
			+94	
	−14	−14	−14	
1st floor	A1	B1	C1	
147 kNm	−49 (−512)	(49) −49	(561) −49	
	63	15	−290	
	212			
	226	−34	−339	Final moments

Note: The net beam connection moments which have to be balanced with the adjacent stanchion moments are $(-M_1 = -512)$, $(561 - 512 = 49)$ and $(M_{pb} = 561)$ kNm respectively. The method does not quite work at C1 where $220 + 339 = 559$ not 561.

assessment must be made of the achievable erected frame tolerances and their compatibility with the wall construction details.

The bending moments in the stanchions are indeterminate unless a computer program is used that can trace the plastic hinge history and give the resulting frame bending moments at the ultimate limit state.

Table 6.6 Plastic moment distribution of stanchion moments for load case 2 (no pattern loading).

Shear moments per storey				
	157	−43	−163	Final moments
	160	−20	−140	
69 kNm	−23	−23	−23	
3rd floor	A3	B3	C3	
	−26 (−335)	(133) −26	(468) −26	
	49		−160	
	175		−119	
	198	−90	−305	Final moments
154 kNm				
	224	−26	−157	Final moments
	250	0	−250	
			+119	
	−26	−26	−26	
2nd floor	A2	B2	C2	
	−37 (−335)	(133) −37	(468) −37	
	63	−70	−78	
	85		−196	
	111	−107	−311	Final moments
222 kNm				
	263	−37	−141	Final moments
	300	0	−300	
			+196	
	−37	−37	−37	
1st floor	A1	B1	C1	
355 kNm	−118 (−335)	(133) −118	(468) −118	
	155	22	−212	
	35			
	72	−98	−330	

Note: The net beam connection moments which have to be balanced with the adjacent stanchion moments are $(-M_1 = -335)$, $(468 - 335 = 133)$ and $(M_{pb} = 468)$ kNm respectively. The method does not quite work at C_1 where $330 + 141 = 471$ not 468.

The simplified design method avoids the use of such programs and bases the analysis on manual calculations supported by linear (first-order) elastic analysis.

The stanchion bending moments in each storey must be in equilibrium with the sway moments caused by horizontal loads such as notional horizontal loads or wind loads. However, with some engineering judgement and using the plastic moment distribution method described in Reference 6.1, reasonable stanchion moments can be derived. A major difference with the Reference 6.1 method is that, here, the beam moments have been determined and that all of the balancing is done within the stanchions. Two examples of the procedure are given in Tables 6.5 and 6.6.

An inspection of Tables 6.3 to 6.6 indicates that the first storey stanchions on lines B and C are critical. The relevant member loads are shown in Table 6.7.

A preliminary choice of stanchion section is $356 \times 368 \times 129$ UC in design grade 43 steel:

The stanchion effective length on its y–y axis is $0.85 \times 5.0 = 4.25$ m.

From *Steelwork Design*[6.2], the data for a stability check to clause 4.8.3.3 are:

Effective length		4 m	5 m	4.25 m
Page 80	P_{cy} kN	3730	3430	3655
Page 135	M_b kNm	659	622	650

From Table 18 of BS 5950, Part 1, for $\beta = 0$, $m = 0.57$.

Hence the unity factor calculations:

$$\frac{0.57 \times 34}{650} + \frac{2593}{3655} = 0.739$$

$$\frac{0.57 \times 339}{650} + \frac{1583}{3655} = 0.731$$

Table 6.7 Critical member loads.

	Axial loads (kN)	Bending moments (kNm)		β
Load case 1 from	2593	34	0	0
Tables 6.3 and 6.5	1583	339	0	0
Load case 2 from	2087	98	0	0
Tables 6.4 and 6.6	1313	330	0	0

$$\frac{0.57 \times 98}{650} + \frac{2087}{3655} = 0.657$$

$$\frac{0.57 \times 330}{650} + \frac{1313}{3655} = 0.649$$

Therefore the preliminary sections are satisfactory.

6.4.4   *Elastic critical load factor λ_{cr} and the required rigid-plastic load factor λ_p*

The structural model and loadings for the consideration of second-order effects using the simplified method are shown in Fig. 6.7.

The elastic critical load factor, λ_{cr}, is estimated in accordance with Appendix F2 of BS 5950. The preliminary section sizes calculated above with pinned bases were found to be inadequate and therefore alternative stanchion sizes with various base spring stiffnesses were analysed using a linear elastic analysis program and the results obtained are summarised in Table 6.8. The following sections were used in these analyses:

Section 1 $356 \times 368 \times 129$ UC Section 2 $356 \times 368 \times 153$ UC
Section 3 $254 \times 254 \times 73$ UC Section 4 $203 \times 203 \times 60$ UC

where, in Table 6.8,

λ_{cr} is always critical in the first storey and equals $\dfrac{1}{200 \times \dfrac{d_1}{5000}} = \dfrac{25}{d_1} \geq 4.6$

$$\lambda_p = \frac{0.9 \times \lambda_{cr}}{(\lambda_{cr} - 1)} \quad \text{if } \lambda_{cr} \leq 10, \quad \text{otherwise } \lambda_p = 1.0$$

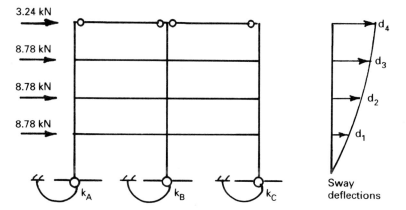

Fig. 6.7 Structural model for simplified second-order analysis.

$$k_p = 10\% \left(\frac{4EI}{h}\right) \quad \text{for a nominally pinned base, and}$$

$$k_p = 100\% \left(\frac{4EI}{h}\right) \quad \text{for a nominally rigid base.}$$

Inf = Infinity, i.e. a fully fixed base.

These values for k_p are in accordance with Reference 6.3. The conclusions from the trial analyses may be summarised as follows:

Trial 1   sections are inadequate
Trial 2   sections are adequate with fully fixed bases. However, this requires large expensive bases and foundations, and also stanchion splices
Trial 3   sections are inadequate
Trial 4   sections are adequate but require a base fixity larger than that recommended for a nominally pinned base
Trial 5   sections are adequate with nominally rigid bases and thus require large expensive bases and foundations
Trial 6   sections are inadequate
Trial 7   sections are just about acceptable with the nominal stiffness of pinned bases.

It follows that the satisfaction of the requirements for second-order effects in sway frames is a non-trivial exercise and may require a significant increase in the section sizes of the stanchions.

Table 6.8 Summary of trial analyses to satisfy second-order stability criteria.

Trial No.	Stanchions				Base stiff. k_p used	d_1 (mm)	d_2 (mm)	d_3 (mm)	d_4 (mm)	λ_{cr}	Required λ_p
	Side		Centre								
	Section	k_p	Section	k_p							
1	1,3	0	1,4	0	0	8.1	10.5	12.6	14.6	3.09	No good
2	1,3	Inf	1,4	Inf	Inf	2.1	3.8	5.8	7.8	11.96	1.0
3	1	6 609	1	6 609	6 609	6.2	8.3	9.5	10.4	4.02	No good
4	1		1		12 000	5.37	7.4	8.5	9.5	4.66	1.146
5	1	66 090	1	66 090	66 090	3.17	4.9	6.0	7.0	7.88	1.031
6	1	6 609	2	8 000	6 609 +8 000	5.88	7.9	9.0	9.9	4.25	No good
7	2	8 000	2	8 000	8 000	5.47	7.5	8.6	9.5	4.57 < 4.6 but accept	1.152

As a result of the above analyses, the preliminary stanchion section size chosen on the basis of a first-order (rigid-plastic) analysis needs to be increased and the required steel section sizes now are:

All beams $\quad\quad$ 533 × 210 × 92 UB in design grade 43
All stanchions $\quad$ 356 × 368 × 153 UB in design grade 43

The stanchion bases are nominally pinned but a notional base stiffness of 10% of the column stiffness according to BS 5950, clause 5.1.2.4 is necessary in order to ensure stability.

Although there is no increase in the preliminary beam size, the original choice gave a rigid-plastic load factor of 1.16 which is greater than the required λ_p of 1.152. The new design is therefore adequate with regard to second-order effects.

6.4.5 Final check of the sway frame at the ultimate limit state

Having modified the design in order to accommodate second-order effects, it is now necessary to carry out a final check of member stability *under the bending moments associated with the required increase in λ_p.*

The loading combinations for this check are again:

Case 1 $\quad$ 1.4 × Dead + 1.6 × Imposed (not pattern) + Notional loads
Case 2 $\quad$ 1.2 × Dead + 1.2 × Imposed (not pattern) + 1.2 × Wind.

The detailed calculation of bending moments given in Section 6.4.3 can be reused here. It is merely necessary to increase the moments by a factor of 1.152. The checks required are for beam and stanchion capacity and stanchion stability. However, the capacity of the beams has effectively been checked above. It may be noted in this respect that the end reaction shear load does not reduce the plastic moment of resistance of the beams.

The member loads for the stanchion checks are shown in Table 6.9.

Table 6.9 Member loads for stanchion checks.

	Axial loads (kN)	Bending moments (kNm)		β
Load case 1 from	2593	39	0	
Tables 6.3 and 6.5	1583	391	0	0
				0
Load case 2 from	2087	113	0	0
Tables 6.4 and 6.6	1313	380	0	0

From *Steelwork Design*[6.2], the data for an overall buckling check to clause 4.8.3.3 of BS 5950 are:

$356 \times 368 \times 153$ UC in design grade 43 steel:

		4 m	5 m	4.25 m (as before)
Effective length				
Page 80	P_{cy} kN	4430	4070	4340
Page 135	M_b kNm	787	751	778

Hence the unity factor calculations with $\beta = 0$, $m = 0.57$ from Table 18 of BS 5950: Part 1, as before:

$$\frac{0.57 \times 39}{778} + \frac{2593}{4340} = 0.626$$

$$\frac{0.57 \times 391}{778} + \frac{1583}{4340} = 0.652$$

$$\frac{0.57 \times 113}{778} + \frac{2087}{4340} = 0.564$$

$$\frac{0.57 \times 380}{778} + \frac{1313}{4340} = 0.581$$

The stanchions also have to be checked for adequate *local capacity* and the data necessary for a check according to BS 5950, clause 4.8.3.2(a) is available from page 176 of *Steelwork Design*[6.2], thus:

$$A_g p_y = 5180 \text{ kN} \quad \text{and} \quad M_{cx} = 787 \text{ kNm}.$$

The required unity factor calculations are therefore:

$$\frac{39}{787} + \frac{2593}{5180} = 0.550 \qquad \frac{391}{787} + \frac{1583}{5180} = 0.803$$

$$\frac{113}{787} + \frac{2087}{5180} = 0.546 \qquad \frac{380}{787} + \frac{1313}{5180} = 0.736$$

and the stanchions are satisfactory with regard to both local capacity and overall buckling. These factors have sufficient excess capacity to allow for holes in the stanchion flanges.

6.4.6   Final check of the non-sway frame at the ultimate limit state

This check introduces pattern loading for the first time and is necessary in order to ensure that stability of the stanchions is achieved and in particular the internal stanchions. The methodology used is relatively unsophisticated and a more precise estimate of the bending moments and axial forces may be obtained if required.

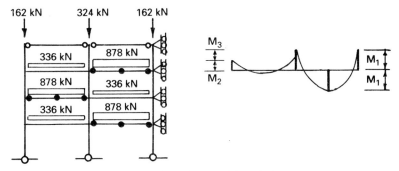

Fig. 6.8 Non-sway pattern loading for stanchion design.

The critical loading combination for this check is $1.4 \times$ Dead $+ 1.6 \times$ Imposed (including pattern loading) and a suitable loading arrangement is shown in Fig. 6.8 together with an indication of the out of balance bending moments on the centre stanchion at each floor level.

The beam load of 878 kN shown in Fig. 6.8 includes the 6% reduction for beam design and this should be increased to 912 kN when calculating the axial loads in the stanchions. A suitable estimate of these bending moments is as follows:

$$M_1 = 878 \times \frac{10}{16} = 549 \text{ kNm} \qquad \text{(plastic collapse mechanism)}$$

$$M_2 = 336 \times \frac{10}{12} = 280 \text{ kNm} \qquad \text{(elastic bending moment}$$
$$\text{distribution)}$$

$$M_3 = M_1 - M_2 = 269 \text{ kNm} \qquad \text{(out of balance moment at each}$$
$$\text{floor level of the centre stanchion)}$$

A suitable model for the redistribution of M_3 into the centre stanchion and the elastic floor beams is shown in Fig. 6.9.

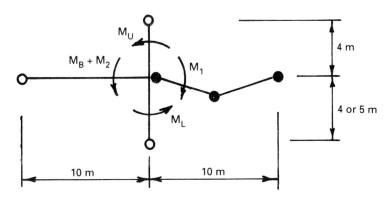

Fig. 6.9 Model for the redistribution of the out of balance moment.

Table 6.10 Moment distribution.

Floor	K_B	K_U	K_L	$\sum K$	M_B (kNm)	M_U (kNm)	M_L (kNm)
3	$\dfrac{55\,330}{10\,000} = 5.53$	$\dfrac{48\,640}{4\,000} = 12.16$	12.16	29.85	49	110	110
2	5.53	12.16	12.16	29.85	49	110	110
1	5.53	12.16	$\dfrac{48\,640}{5\,000} = 9.73$	27.42	54	119	96

In Fig. 6.9, M_B, M_U and M_L are the changes in bending moment necessary to bring the joint into equilibrium. These are proportioned according to their individual stiffnesses, $K = I/L$ as shown in Table 6.10, so that

$$M_i = \frac{K_i M_3}{K_B + K_U + K_L} \quad \text{where } K_i = K_B, \ K_U \text{ or } K_L$$

and $I_B = 55\,330 \text{ cm}^4$, $I_C = 48\,640 \text{ cm}^4$.

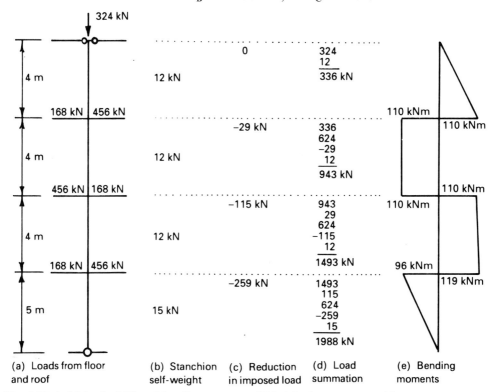

(a) Loads from floor and roof

(b) Stanchion self-weight

(c) Reduction in imposed load

(d) Load summation

(e) Bending moments

Fig. 6.10 Axial loads (kN) and bending moments (kNm) in the centre stanchion.

Hence the centre stanchion bending moments and axial loads are summarised in Fig. 6.10.

The stability calculations are carried out in accordance with clause 4.8.3.3.1 of BS 5950 and the required data is available in *Steelwork Design* [6.2]:

$356 \times 368 \times 153$ UC design grade 43 steel

		3.0 m	4.0 m	5.0 m	3.4 m	4.25 m
Effective length						
Page 80	P_{cy} kN	4760	4430	4070	4628	4340
Page 135	M_b kNm	787	787	751	787	785

1st storey $M_A = 96$ kNm with $\beta = 0$ and $m = 0.57$, $F = 1988$ kN.

$$\frac{0.57 \times 96}{785} + \frac{1988}{4340} = 0.528 < 1.0$$

2nd storey $M_A = 119$ kNm with $\beta = 110/119 = 0.9244$ and $m = 0.96$, $F = 1493$ kN.

$$\frac{0.96 \times 119}{787} + \frac{1493}{4628} = 0.468 < 1.0$$

and therefore the stanchions are satisfactory.

6.4.7 Elastic analysis check under unfactored loads

The final check that is required in order to confirm the member selection using the simplified method is given in clause 5.7.3.3 (c) of BS 5950, Part 1. This states that:

'Under all (reasonable) combinations of *unfactored* loading (including the notional horizontal loads when wind loads are not included in the combination) it should be possible by means of moment re-distribution to produce sets of moments and forces throughout the frame which are in equilibrium with the applied loads and under which all members remain elastic.'

The background to this clause is given in Section 3.1.3 of this book. It is evidently intended to provide a simplified procedure to ensure that the structure has the capacity to *shake down* under the unfactored loads thus avoiding either alternating plasticity or incremental collapse under repeated fluctuations of load.

The wording of the above clause appears to be imprecise and liable to cause confusion. No limit is placed on the amount of redistribution of bending moment that is allowed and therefore it does not exclude using plastic collapse distributions similar to those that were the basis of the primary design calculations. On the face of it, bearing in mind that the design uses load factors greater than 1.2 for all loading cases and that UB and UC members have shape factors of the order of 1.15, satisfaction of this clause would appear to be a non-event.

If this clause is intended to provide a shakedown check, then logically *precisely the same redistribution should be applied to all loading cases.* The clause does not appear to require this but the authors would suggest this as a reasonable interpretation of its requirements. This being the case, the easiest way to satisfy this clause is to carry out the necessary

Table 6.11 Bending moments 'M' and axial loads 'F' in stanchions.

Stanchion/ floor level	Dead + Imposed (NP)		Dead + Imposed (P)		Dead + Imposed (NP) + Notional Loads		Dead + Imposed (NP) + Wind		Dead + Imposed (P) + Notional Loads	
	M (kNm)	F (kN)	M	F	M	F	M	F	M	F
A1L	135	838	36	508	111	830	64	808	18	502
1U	260	562	125	399	259	558	254	546	124	396
2L	219		218		209		186		210	
2U	217	279	218	111	215	277	205	272	217	110
3L	253		125		247		233		121	
3U	153		43		151		138		41	
B1L	0	1795	115	1270	31	1795	93	1795	92	1270
1U	0	1191	*	843	13	1191	46	1191	*	843
2L	0		86		19		60		100	
2U	0	600	91	425	7	600	28	600	96	425
3L	0		*		10		38		*	
3U	0		132		5		27		129	
C1L	135	838	156	678	159	847	206	868	173	684
1U	260	562	244	396	261	566	267	577	244	399
2L	219		93		230		253		101	
2U	217	279	89	284	219	280	229	286	91	285
3L	252		233		258		272		237	
3U	153		175		156		169		176	

* = smaller moment and giving single curvature in member
P = Pattern loading
NP = Non-pattern loading
U = stanchion level just above floor beam
L = stanchion level just below floor beam
Additional axial loads in the stanchions are given in Table 6.12.

elastic analyses (by computer if possible) and then to carry out any redistribution manually if this is required. In general, a properly designed frame should be capable of sustaining the unfactored loads elastically without the need for significant redistribution.

Bending moments and axial loads from linear elastic computer analyses are summarised in Tables 6.11, 6.12 and 6.13. A capacity check to clause 4.8.3.2(a) of BS 5950 is used to check the stanchions for elastic action and if a member fails the clause requirements then moment redistribution will have to be considered. From Section 6.4.5 above, $A_g p_y = 5180\,\text{kN}$ and $M_{cx} = 787\,\text{kNm}$.

$$\text{Hence check for } \frac{M}{M_{cx}} + \frac{F}{A_g p_y} \le 1.0$$

By inspection of Tables 6.11 and 6.12 the stanchions are satisfactory, provided that the beams are satisfactory. The beams have a plastic moment of resistance of 651 kNm and this is not exceeded in Table 6.13 and therefore the frame is satisfactory.

6.4.8 Deflections at the serviceability limit state

6.4.8.1 General
In general, the calculation of the deflections of a plastically designed multi-storey structure requires the use of a linear elastic computer

Table 6.12 Stanchion axial loads to be added to those given in Table 6.11.

Stanchion/ floor level	Non-pattern loading				Pattern loading			
	Roof	Self-wt./ Wall	Imposed load red.	Total (kN)	Roof	Self-wt./ Wall	Imposed load red.	Total (kN)
A3U	113	50	0	163	113	50	0	163
A2U/A3L	113	100	−7	206	113	100	0	213
A1U/A2L	113	150	−50	213	113	150	−7	256
A1L	113	200	−130	183	113	200	−7	306
B3U	225	8	0	233	225	8	0	233
B2U/B3L	225	16	−14	227	225	16	−7	234
B1U/B2L	225	24	−100	149	225	24	−50	199
B1L	225	34	−260	−1	225	34	−130	129
C3U	113	50	0	163	113	50	0	163
C2U/C3L	113	100	−7	206	113	100	−7	206
C1U/C2L	113	150	−50	213	113	150	−7	256
C1L	113	200	−130	183	113	200	−50	263

Table 6.13 Bending moments in floor beams.

Loading	Floor	Span A–B			Span B–C		
		Line A	Max Sag	Line B	Line B	Max Sag	Line C
D + Imp	3	405	265	513	513	265	405
(NP)	2	436	256	499	499	256	436
	1	395	267	522	522	267	395
D + Imp	3	167	88	262	464	288	407
(P)	2	435	285	442	265	79	182
	1	160	87	272	467	291	399
D + Imp	3	398	266	521	506	265	414
(NP) + NL	2	423	257	512	487	256	449
	1	370	269	544	500	264	419
D + Imp	3	371	268	545	482	263	441
(NP) + W	2	390	259	543	456	255	482
	1	317	276	591	453	261	472
D + Imp	3	161	89	267	459	288	413
(P) + NL	2	426	285	451	256	78	191
	1	141	90	288	451	290	417

analysis. Here, the calculation includes the nominal stiffness of the pinned bases which has been assumed to be 10% of $4EI/h = 8000$ kNm/rad.

6.4.8.2 Wind sway
The maximum horizontal deflections under wind loading are shown in Table 6.14.

As the permissible storey height to deflection ratio is 300, these values are satisfactory.

6.4.8.3 Floor beam deflections
The maximum floor beam deflections under vertical load are shown in Table 6.15.

Table 6.14 Maximum horizontal deflection.

	Sway (mm)	Nett storey (mm)	Storey/deflection
Roof	21.4	3.0	1320
3rd floor	18.4	2.9	1379
2nd floor	15.5	4.4	921
1st floor	11.1	11.1	448

Table 6.15 Maximum vertical deflections.

	mm	Span/deflection
Dead load	7.6	1315
Imposed, non-pattern	10.7	934
Imposed, Pattern	12.7	787
Dead + Imposed, Pattern	20.3	493

As the permissible span to deflection ratio *under imposed load alone* is 360, these values are satisfactory.

6.4.9   Design of the floor beam to stanchion connection

6.4.9.1   General

This connection is at a plastic hinge location and therefore should be designed to have a satisfactory moment rotation capacity or it should be over-designed in order to ensure that the plastic hinge occurs only in the beam member. Moment-rotation capacity would not be required if the connection was only at the last hinge to form or if it could be shown that plastic hinges at the connections only occurred at levels of loading above the required factored loading (enhanced by λ_p according to Section 6.4.4).

In this example the connection will be over-designed by a factor of 1.2 in order to ensure that any over-strength steel in the beam does not force the plastic hinge into the connection.

The plastic moment of resistance of the beam $= 651$ kNm, hence the connection will be designed to resist a moment of $1.2 \times 651 = 781$ kNm.

The design vertical shear load on the connection is

$$(1.4 \times 24 + 1.6 \times 33.84) \times 4.82 \times 1.2 = 508 \text{ kN}.$$

The connection can be either site welded or site bolted. A site welded connection will keep the connection within the depth of the beam with no projection of material beyond the stanchion flange.

A site bolted connection will need to be deeper than the beam depth in order to achieve a lever arm long enough to develop sufficient bending moment from the upper bolt group. The connection will also need to project outside the stanchion flanges for up to 300 mm so that the bolt loads and compression zone loads can be transferred gradually into the beam. The additional material in the connection will raise the beam strength above its basic plastic moment of resistance locally but this should not be of any concern because additional strength will not weaken the structure. The stanchion will not become the weak link in the structure because the depth of the beam connection also reduces the bending moments that occur in the stanchion. It may be noted that the finite lengths and depths of haunches are relatively easily taken into account in the plastic design of portal frames whereas the lengths and depths of beam to stanchion connections in multi-storey frames cannot be included without using computer programs to carry out the necessary number processing.

6.4.9.2 Connection design example

A preliminary design for a site bolted connection is shown in Fig. 6.11. The details of such a connection should be discussed with the fabricator before the design is finalised.

There are four pairs of bolts in the top bolt group and two pairs in the bottom group. The triangular stiffener on the top flange of the beam dissipates the bolt tensile loads into the beam flange. A pair of smaller triangular stiffeners are added to this larger stiffener in order to

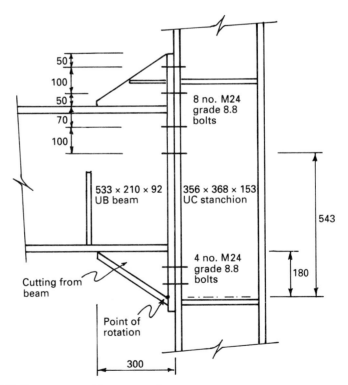

Fig. 6.11 Beam to stanchion connection.

strengthen the end-plate and to reduce the prying action of the bolts. The latter effectively increases the allowable direct tensile loads in the bolts and hence the moment capacity of the connection. However, in this example the allowable bolt loads have been limited to the BS 5950 standard values. The bottom stiffener is a cutting from the beam section. The sizes of these three stiffeners are chosen such that they will lie within the depth of the concrete floor.

Bolts. The upper bolt group provides the tensile force and has a lever arm from the point of rotation which is taken to be adjacent to the flange of the lower stiffener as shown in Fig. 6.11. Assuming that each bolt has a tensile capacity T kN and that only the top two pairs of bolts are fully loaded, then the moment capacity equals:

$$2T\left(0.863 + 0.763 + \frac{0.643^2}{0.763} + \frac{0.543^2}{0.763}\right) = 5.108T \text{ kNm} = 781 \text{ kNm}$$

$$\therefore \quad T = 153 \text{ kN}$$

Therefore, use M24 grade 8.8 bolts. From page 244 of Reference 6.2, the tensile capacity $= 159 \text{ kN/bolt}$.

Beam top flange stiffener. The part of the tensile load which is resisted by 100 mm depth of the stiffener is approximately 2×153 kN. Use a 12 mm thick stiffener with 10 mm fillet welds.

End-plate. The end-plate is 220 mm wide with bolts at 120 mm horizontal cross centres. From the formula used previously in Section 5.9.4 of Chapter 5:

$$t = \sqrt{\frac{F_t m}{p_y L_e}}$$

with $\quad F_t = 4 \times 153 = 612 \text{ kN}$

$$m = \left(\frac{120 - 12}{2}\right) - 10 = 44 \text{ mm}$$

which gives $\quad t = \sqrt{\dfrac{612 \times 44}{0.265 \times 100}} = 31.9 \text{ mm}$

As this is impracticably thick, use two triangular stiffeners, $90 \times 10 \times 150$, to strengthen the end-plate as shown in Fig. 6.11 and hence use a 20 m thick plate. Strictly speaking, this should be justified, for instance by yield line analysis. However, here it will be adopted without further justification. Similarly, stiffeners are also necessary to strengthen the 20.7 mm thick flange of the stanchion where it too has to resist the bolt tensile forces.

Shear load capacity of the lower bolt group.
Design shear force = 508 kN.
Shear capacity of M24 grade 8.8 bolt = 132 kN/bolt.
Bearing capacity on 20 mm grade 43 steel plate = 221 kN/bolt.
Therefore the capacity of the four bolt group is $4 \times 132 = 528$ kN which is adequate.

Note: There is also some shear load capacity from the upper bolt group.

Capacity of cutting used to stiffen the lower flange.

$$\text{Horizontal compressive force} = 2 \times 153 \left(2 + \frac{0.643}{0.763} + \frac{0.543}{0.763} \right)$$

$$= 1087 \text{ kN}$$

$$\text{Flange force} = 1087 \frac{\sqrt{180^2 + 300^2}}{300} = 1268 \text{ kN}$$

A cutting from the beam section has a flange compression capacity of $15.6 \times 209.3 \times 0.275 = 898$ kN. Hence the web of the cutting takes $1268 - 898 = 370$ kN. Theoretically, for the point of rotation to be correct, up to about 50% of the compression load could be taken by the web.

Shear capacity of stanchion web.
Design bending moment = 781 kNm
Lever arm = 781/1087 = 0.718 m   (1087 kN horizontal force
                                                                                  – see above)

Design shear force = 781/0.718 = 1088 kN
Design shear strength, $P_y = 725$ kN   (Reference 6.2, page 141)

The bending capacity of the stanchion is reduced if the shear force is greater than $0.6P_y$. Therefore 'K' type web shear stiffeners are required to take the excess shear force of $1088 - 0.6 \times 725 = 653$ kN. The area of the required stiffeners is therefore

$$\frac{653\,000 \times (\text{say}) \, 1.5}{275} = 3562 \text{ mm}^2$$

Therefore use 2 No. 15×150 stiffeners.

Note: The connection design given above is simple and conventional. A more rigorous approach is given in Reference 6.4.

References

6.1   Horne M R & Morris L J. Plastic design of low-rise frames. Constrado Monograph. William Collins, 1981.
6.2   The Steel Construction Institute. Steelwork Design: Guide to BS 5950: Part 1: 1990, Volume 1 – Section properties and member capacities, 1992.
6.3   Advisory Desk Note AD 097. Nominal base stiffness. *Steel Construction Today*, November 1991.
6.4   The Steel Construction Institute/The British Constructional Steelwork Association. Joints in steel construction – Moment connections, Publication No. 207/95, 1995.

Chapter 7
Miscellaneous Portals

7.1 Introduction

Chapter 2 introduced the reader to the general principles of plastic design and Chapter 5 described the detailed design of a single-bay building. Many other shapes of portal frame not considered in these earlier chapters can readily be designed using plastic theory. This chapter extends the earlier ideas to a wide range of different shapes of portal frame. Attention is confined to deriving the bending moment diagram at collapse and choosing suitable member sizes. Verifying member stability and designing the connections follows the same principles that have been described in the earlier chapters and is not considered further here.

This chapter concentrates on graphical and semi-graphical analyses using bending moment diagrams drawn to scale. These methods have great advantages for the type of frames being considered here in locating the plastic hinges and 'seeing' the collapse mechanism. They have great versatility and can be applied to a wide variety of frames and load cases. Mathematical verification of a diagram can be used afterwards to improve its accuracy if this is considered necessary.

As described in Chapter 2, there are two methods available for pinned based frames. The first cuts the frame at the apex and introduces three redundancies. For this method the reactant bending moment diagram can then be drawn using the two necessary and sufficient geometric conditions derived in Section 2.6.4 or by solving simultaneous equations for equilibrium at the hinge positions. Alternatively, the frame can be made statically determinate by releasing the horizontal restraint at one base. The reactant bending moment diagram can then be drawn very easily in terms of the single redundant horizontal force.

For fixed based frames, only the first of these methods is available. However, the provision of fixed bases for single-storey portal frames is extremely expensive and it is usually found to be more economical to use pinned bases despite the extra material required in the frames. Both methods are then available and in many cases the choice is largely a matter of personal preference. In the first method, the drawing of the 'free' bending moment diagram is much simpler but the manipulation of the reactant moment diagram rather more complicated. Conversely, with the second method, more effort may be required to draw the free bending moment diagram but then the reactant line can be drawn and adjusted very easily.

Both of these methods will be used in this chapter but the authors would not claim that they have necessarily used the best method in each case.

When using the first method, some of the portal frames considered here, such as tied portals and northlight portals, have a reactant bending moment diagram which follows the free diagram quite closely. The actual bending moment at any section is then the relatively small difference between two larger quantities. In such cases, a graphical analysis based on scaled bending moments should always be checked mathematically in order to eliminate the magnification of drawing inaccuracies. It is also doubly important to ensure that all reasonable load combinations are considered.

The remainder of this chapter will illustrate these points by a series of examples.

7.2   Crane buildings

The dynamic loads and impact factors to be used in the design of crane buildings are specified in BS 5950: Part 1, clause 2.2.3. This clause simply refers the reader to BS 6399: Part 1. It also recommends that values for cranes of loading class Q3 and Q4 as defined in BS 2573: Part 1 should be established in consultation with the crane manufacturer.

Crane beams and other members directly supporting cranes of utilisation classes U4 to U9 as defined in BS 2573: Part 1 need to be checked for fatigue. Clause 2.4.3 of BS 5950 refers the reader to BS 5400: Part 10.

The load factors and load combinations for the design of structures supporting cranes are given in BS 5950: Part 1, clause 2.4.1.2 and Table 2. The serviceability limit state specified in clause 2.5.1 states that crane surge and wind need not be considered together in any loading combination. Deflection limits are given for individual crane gantry girders but nothing is specified for the variations in the horizontal distances between the crane rails. Designers should build up their experience by learning from actual built designs and from discussions with crane manufacturers. In the absence of any other guidance, the following values can be used for cranes of loading classes Q1 and Q2.

The transverse horizontal deflection at the top of the column $\leq h/300$.

The theoretical spread between the crane rails at any point along the rails due to reasonable combinations of frame loads and crane loads $\leq \pm20\,\text{mm}$.

Fixed bases are beneficial in reducing deflections but they can considerably increase the cost of the foundations especially when building on filled sites. If nominally pinned bases are used, then consideration should be given to taking advantage of the advice in clause 5.1.2.4(b) for nominal foundation stiffness. However, this should be used cautiously and should probably be limited to buildings where the cranes are of utilisation classes U1 to U3.

7.2.1   Example 7.1. Design of a crane building

The dimensions of the building and the required loads at the ultimate limit state are shown in Fig. 7.1.

   The frame is cut at the apex and three redundancies, *M*, *H* and *V* introduced as shown in Fig. 2.58. The resulting free and reactant bending moment diagrams are shown superimposed in Fig. 7.2.

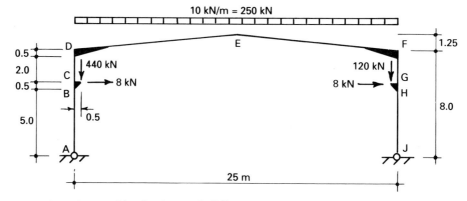

Fig. 7.1   Dimensions and loads of crane building.

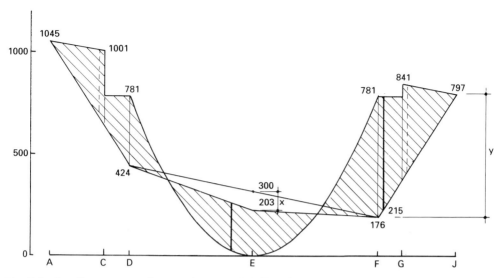

Fig. 7.2   Bending moment diagram for crane building (kNm).

The free bending moment diagram is drawn as follows.

At E $\qquad = 0$

E to D and E to F $\qquad = 5x^2$ $\qquad$ with x measured from E in plan

D to C and F to G $\qquad = 781.25 \, \text{kNm}$ $\qquad$ constant

Below C $= 781.25 + 220 = 1001.25 \, \text{kNm}$

At A $= 1001.25 + 8 \times 5.5 = 1045.25 \, \text{kNm}$ $\qquad$ with A to C a straight line

Below G $= 781.25 + 60 = 841.25 \, \text{kNm}$

At J $= 841.25 - 8 \times 5.5 = 797.25 \, \text{kNm}$ $\qquad$ with G to J a straight line

In order to draw the reactant line, we need to make an assumption and it is usually easier to start by guessing a member size for the column. We will choose a $533 \times 210 \times 82 \, \text{UB}$ in Grade 43 steel with a full plastic moment of 566 kNm. Inspection (or a trial reactant line if this is not sufficient) shows that the critical column bending moment is below the haunch at F. Fixing the actual bending moment (the difference between the free and reactant values) at this point $= 566 \, \text{kNm}$ allows the reactant line between A and F to be drawn with a value below the haunch of $781 - 566 = 215 \, \text{kNm}$ and a value at F of 176 kNm.

As shown in Section 2.6.4, the ordinate of the reactant diagram for the left-hand column between A and D must be the same as that between J and F. This ordinate is 'y' $= 797 - 176 = 621 \, \text{kNm}$ so that the reactant moment at D is $1045 - 621 = 424\,\text{kNm}$ and the reactant line between A and D can be drawn as shown. The dimension 'x' (see Fig. 2.72) now follows as:

$$x = \frac{ay}{b} = \frac{1.25 \times 621}{8} = 97\,\text{kNm}$$

The remainder of the reactant line can now be drawn and the maximum bending moment in the rafter scaled at about 220 kNm. As here the scale drawing has been repeated numerically, it is easy to obtain an exact value of the maximum rafter moment by calculation. In the vicinity of the rafter hinge, with x measured horizontally from E, the equation of the free bending moment is

$$M_{ED} = -5x^2$$

and the equation of the reactant bending moment is

$$M_{ED} = 203 + (424 - 203)\,\frac{x}{12.5}$$

giving for the shaded region in Fig. 7.2

$$M_{ED} = 203 + 17.68x - 5x^2$$

This has a maximum value when

$$\frac{dM}{dx} = 0 \quad \text{i.e.} \quad x = 1.768\,\text{m}$$

so that, substituting this value of x into the equation for M_{ED}, the maximum bending moment in the rafter is

$$M_{max} = 203 + 31.3 - 15.6 = 218.7\,\text{kNm}$$

There are several suitable rafter sections and the lightest is a $406 \times 140 \times 46\,\text{UB}$ with $M_p = 245\,\text{kNm}$. However, second-order effects are likely to be significant here and there is no virtue in being too miserly. For the moment, therefore, we will select a $457 \times 152 \times 52\,\text{UB}$ with $M_p = 301\,\text{kNm}$.

It is instructive to compare the two available graphical methods and, for this frame, the second method proceeds as follows. Releasing the horizontal force at the right-hand base makes the frame statically determinate and the vertical reactions V_A and V_J at A and J respectively are determined first. By equilibrium of moments about A:

$$25V_J = 250 \times 12.5 + 120 \times 24.5 + 440 \times 0.5 + 16 \times 5.5$$

$$= 6373\,\text{kNm}$$

$$\therefore \quad V_J = 254.9\,\text{kN}$$

$$V_A = 250 + 560 - 254.9 = 555.1\,\text{kN}$$

The total applied horizontal load of $16\,\text{kN}$ is all resisted at A and the free bending moments can now be calculated. Between A and C with x measured from A to C

$$M_{AC} = 16x$$

Between C and D with x still measured from A

$$M_{CD} = 16x - 8(x - 5.5) + 220 = 264 + 8x\,\text{kNm}$$

Thus, in the left-hand stanchion

$M_C = 88\,\text{kNm}$ below the crane rail and $308\,\text{kNm}$ above it
$M_D = 328\,\text{kNm}$

Between D and E with x measured from D to E in plan

$$M_{DE} = (555.1 - 440)x + 16\left(8 + \frac{1.25x}{12.5}\right) - 8\left(2.5 + \frac{1.25x}{12.5}\right)$$

$$+ 220 - \frac{10x^2}{2}$$

$$= 328 + 115.9x - 5x^2\,\text{kNm}$$

Thus, in the left-hand rafter

$$M_E = 995\,\text{kNm}$$

Between J and G the free bending moment is zero. Between G and F, with x measured from G

$$M_{GF} = 60 + 8x$$

Thus, in the right-hand stanchion

$M_G = 0$ below the crane rail and $60\,\text{kNm}$ above it
$M_F = 80\,\text{kNm}$

Between F and E, with x measured from F to E in plan

$$M_{FE} = (254.9 - 120)x + 8\left(2.5 + \frac{1.25x}{12.5}\right) + 60 - \frac{10x^2}{2}$$

$$= 80 + 135.7x - 5x^2\,\text{kNm}$$

Thus, in the right-hand rafter

$$M_E = 995\,\text{kNm}$$

The free bending moment diagram obtained in this way is drawn on the same diagram as the reactant diagram in Fig. 7.3.

The reactant bending moment diagram is now constructed very easily. If we start, as before, by assuming a plastic hinge with moment of resistance 566 kNm below the haunch at F, the horizontal reaction H at J is given by

$$7.5H = 76 + 566$$

$$\text{giving} \quad H = \frac{642}{7.5} = 85.6 \, \text{kN}$$

The ordinates of the reactant line are then given by

$$\begin{aligned} M_D = M_F &= 8 \times 85.6 &= 685 \, \text{kNm} \\ M_E &= 9.25 \times 85.6 = 792 \, \text{kNm} \end{aligned}$$

and the reactant line can be drawn as shown. The maximum bending moment in the rafter scales about 220 kNm as before.

The actual load factor against collapse can also be estimated. At the plastic hinge position in the rafter, the magnitude of the free bending moment is 996 kNm and that of the reactant bending moment is 777 kNm. If the moment capacity of the chosen rafter (301 kNm) is added to the reactant moment, as plotted on Fig. 7.3, the free moment capacity available is $777 + 301 = 1078 \, \text{kNm}$. The ratio $1078/996 = 1.08$ provides a good estimate of the actual load factor and shows that there is about 8% overcapacity available to take account of second-order effects.

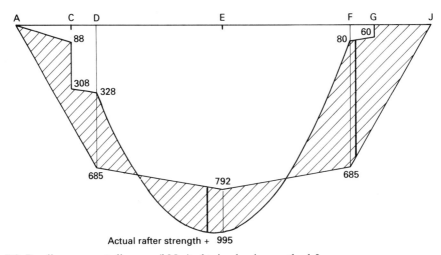

Fig. 7.3 Bending moment diagram (kNm) obtained using method 2.

With this method it is easy to investigate whether an alternative design may be better. The next lightest section to try for the stanchion would be a $457 \times 191 \times 74\,\text{UB}$ with $M_p = 456\,\text{kNm}$. This gives

$$H = \frac{(76 + 456)}{7.5} = 70.9\,\text{kNm}$$

and reactant line ordinates of

$$\begin{aligned}
M_D = M_F &= 8 \times 70.9 &= 567.5\,\text{kNm} \\
M_E &= 9.25 \times 70.9 = 656\,\text{kNm}
\end{aligned}$$

With the reactant line drawn in this way, the maximum bending moment in the rafter scales about $340\,\text{kNm}$ which requires at least a $457 \times 152 \times 60\,\text{UB}$ with $M_p = 353\,\text{kNm}$. However, this would give no reserve of strength to counter second-order effects and it is necessary to anticipate a significant increase in rafter section. The rafter size would then become close to or even equal to the stanchion section and experience suggests that a uniform or near uniform frame is unlikely to be economic for a structure of this type. We will therefore continue with the original design using a $533 \times 210 \times 82\,\text{UB}$ for the stanchion and a $457 \times 152 \times 52\,\text{UB}$ for the rafter.

This example illustrates perfectly the relative merits of the two alternative graphical methods. The two bending moment diagrams shown shaded in Figs 7.2 and 7.3 are the same but the routes to them are quite different. In Fig. 7.2, drawing the free bending moment diagram was relatively easy but the reactant line is rather more difficult to manipulate. In Fig. 7.3, most of the effort went into drawing the free bending moment diagram but once this had been done any number of reactant lines could be tried with little effort.

There are two further tasks to carry out before the choice of member sizes is complete. The first is to check the stanchions for the effect of the rather large axial loads added at B and H. At position B, adding a further $10\,\text{kN}$ (say) of axial load for the weight of the side cladding and noting that the cross-sectional area of a $533 \times 210 \times 82\,\text{UB}$ is $10\,500\,\text{mm}^2$,

$$\text{Bending moment} = 5 \times 85.6 - 88 \times \frac{5}{5.5} = 348\,\text{kNm}$$

$$\text{Axial load} = 555 + 10 = 565\,\text{kN}$$

$$\therefore \quad n = \frac{p}{Y_s} = \frac{565\,000}{10\,500 \times 275} = 0.196$$

$$\text{and} \quad M_p' = 0.275(2058 - 2853n^2) = 536\,\text{kNm}$$

where the formula for the reduced plastic moment M_p' is found on page 30 of Reference 7.1. As the reduced capacity of 536 kNm is greater than the bending moment of 348 kNm, the stanchion is satisfactory.

Similarly, at position H,

$$\text{Bending moment} = 5 \times 85.6 = 428\,\text{kNm}$$

$$\text{Axial load} = 255 + 10 = 265\,\text{kN}$$

$$\therefore \quad n = \frac{p}{Y_s} = \frac{265\,000}{10\,500 \times 275} = 0.091$$

$$\text{and} \quad M_p' = 0.275(2058 - 2853n^2) = 559\,\text{kNm}$$

As the reduced capacity of 559 kNm is greater than the bending moment of 428 kNm, the stanchion is satisfactory.

The second remaining task is to check for second-order effects. However, this is problematical.

The effect of the crane is to increase the sizes of the steel sections compared with a craneless building. However, the crane also increases the axial loads in both the columns and the rafters and particularly in the former. In general, the increase of axial loads has a greater effect than the increase in section sizes so that it is essential to check for second-order effects including the crane loads.

In principle, the formulae given in Section 3.2.3 of this book can be used to check the frame for second-order effects provided that some conservative assumptions are made. Thus the axial loads at collapse arise directly from the plastic design as shown in Fig. 7.4.

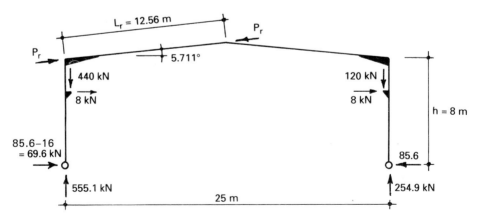

Fig. 7.4 Axial loads at collapse.

As indicated in Chapter 3, in order to use the simplified formulae it is necessary to ignore the beneficial effect of the haunches. It is also necessary to make some assumptions regarding the axial loads in the stanchions and it is conservative to assume that the high axial load below the crane support exists over the whole length of the stanchion. Finally, the two halves of the structure are not equally loaded and it is necessary to apply the formula to the most highly loaded half of the frame. In this case, it is the left-hand side that is most critical and, for this half of the frame, the calculation proceeds as follows:

I_c = Second moment of area of column = 4.752×10^8
I_r = Second moment of area of rafter = 2.137×10^8
P_c = Axial force in column = 551 kN
P_r = Axial force in rafter = $77.6 \cos 5.711° + 115.1 \sin 5.711°$
= 88.7 kN

$$R = \frac{I_c L_r}{I_r h} = \frac{4.752 \times 10^8 \times 12.56}{2.137 \times 10^8 \times 8} = 3.492$$

$$\lambda_{cr} = \frac{3EI_r}{L_r\left[\left(1 + \dfrac{1.2}{R}\right)P_c h + 0.3 P_r L_r\right]}$$

$$= \frac{3 \times 205 \times 2.137 \times 10^8}{12\,560\left[\left(1 + \dfrac{1.2}{3.492}\right)\times 555.1 \times 8000 + 0.3 \times 88.7 \times 12\,560\right]}$$

$$= 1.66$$

This is, of course, an extremely low value and well below the threshold of 4.6 for the use of an approximate treatment of second-order effects as discussed in Section 3.2.1. Certainly, the calculation contains a number of safe assumptions and may therefore be regarded as primarily indicative. However, an 'exact' analysis for the elastic critical load of this frame gives $\lambda_{cr} = 3.43$ which is still too low to continue with an approximate calculation. There are therefore three alternatives:

❏ Increase the member sizes until an elastic critical load factor greater than 4.6 can be reasonably justified (e.g. by relaxing one or more of the safe assumptions such as by taking account of the rotational stiffness of the nominally pinned bases, averaging the axial load in the column over its whole length or sharing the second-order destabilising effect of the crane between the two columns).

❏ Proceed on the basis of a second-order elastic-plastic analysis. For this frame and loading, such an analysis reveals that the failure load is $\lambda_f = 0.93$ so that the increase of section sizes to achieve security in this way is not great.

❏ Revert to elastic analysis.

The writers had not anticipated this problem when they started to write this section on crane buildings and, to the best of their knowledge, the low elastic critical loads of portals supporting cranes have not been discussed previously. This is evidently a subject worthy of further study and for the present it is only possible to point out the problem and to suggest caution when designing this type of structure, especially if the deflection calculation reveals that it is at all flexible.

7.3   Portal frame with an intermediate floor

This is a common type of structure but the advantages offered by the floor are not always recognised and used. A typical situation is shown in Fig. 7.5. Even though the floor beam is only nominally pinned to the stanchions, it effectively creates a fixed base portal frame with the fixed base at first floor level. This gives the advantage of:

❏ a reduction in the size, and therefore cost, of the frame members;

❏ double curvature in the upper length of the stanchion with consequent advantages for member stability.

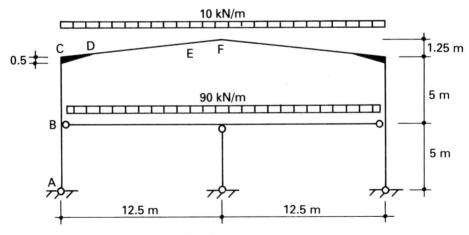

Fig. 7.5 Portal frame with an intermediate floor.

As far as the portal frame is concerned, the floor beam merely serves as a tie between the stanchions. In order to fulfil this function:

☐ the floor beam has to be continuous between the stanchions with respect to tensile forces;
☐ it should have suitable connections to resist the induced tensile forces, as well as the shear forces from the floor loads, without slippage. Any connection slippage or distortion may put some of the tensile force into the concrete floor.

Often, the floor construction will be sufficiently rigid to prevent sidesway of the portal frame at the floor level. When, and only when, this is the case a further significant advantage is obtained:

☐ A larger elastic critical load, making second-order effects insignificant.

It should be appreciated that a different situation is created if the floor beam is provided with a moment connection to the frame. This is not precluded but it should be designed with care as the floor beam may become part of the frame mechanism and thereby reduce the frame strength. This situation should be avoided if the section size of the floor beam is significantly larger than the section size of the stanchions.

The additional bending moment caused by the eccentricity of the connection to the floor beam can be beneficial in reducing the bending moments in the lower storey of the stanchions. However, this effect does depend on the floor beam layout and the resulting eccentricity about the x–x axis of the stanchion. This has not been included in Example 7.2 which follows.

The effective 'base fixity' in the stanchion at the first floor connection will not be equal to the gross plastic moment of the stanchion because of the holes in the column flange for the floor beam connection, and because of the reduction due to the axial load in the stanchion. As far as portal frame behaviour is concerned, there is no benefit in having fixed bases to the foundations. However, fixed bases would improve the stability of the lower length of the stanchion and help to counteract the negative effect of the large axial compressive loads.

7.3.1 *Example 7.2. Portal frame with an intermediate floor*

The frame and its loads at the ultimate limit state are shown in Fig. 7.5. Here, it will be assumed that the floor is able to prevent the frame from swaying at the level of the floor connection. If this was not the case, as far as second-order effects are concerned, the frame would act as though it were a pinned-based frame of 10 m height to eaves and with a large additional axial load over the lower half of the stanchion. Such a frame would be extremely sensitive to second-order effects and a quite different design approach would be required.

Because of the effective base fixity, this frame is best designed by the first of the graphical methods with a cut at the apex. The symmetry of the frame and loading means that there is no vertical redundant force at the cut. However, the floor beam introduces another redundancy T so that the forces contributing to the reactant diagram are as shown in Fig. 7.6. This additional force does not affect the geometry of the reactant diagram above B and it is this part of the diagram that controls the design.

Because of the symmetry, it is only necessary to consider half of the frame. In order to draw the free bending moment diagram, we note that the axial force in the floor beam is released because this force is treated as a redundancy. The free bending moments are therefore exactly as for a portal frame without an intermediate floor as shown in Fig. 7.7, rising in a parabolic curve from zero at F to $10 \times 25^2/8 = 781.25$ kNm at C and then remaining constant at this value from C to A.

The geometric constraints on the reactant diagram which were derived in Section 2.6.4 remain valid. In order to commence the design, we first guess a section size for the stanchion and a $457 \times 152 \times 60$ UB with a full plastic moment of 353 kNm will be tried. As a conservative estimate, this value of M_p will be reduced to $0.75 \times 353 = 265$ kNm above B for the reasons outlined above. If a plastic hinge is also assumed below the haunch, the first part of the reactant diagram between B and C can be drawn as shown in Fig. 7.7.

$$M_A = 781 + 265 = 1046 \text{ kNm}$$
$$M'_C = 781 - 353 = 428 \text{ kNm}$$

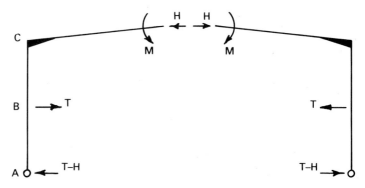

Fig. 7.6 Redundant forces.

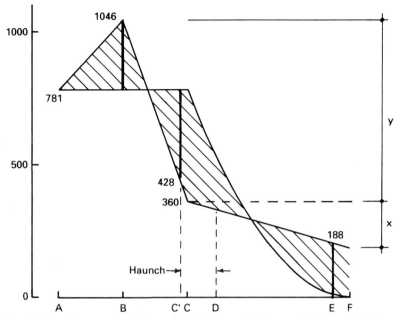

Fig. 7.7 Bending moment diagram for frame with intermediate floor (kNm).

The value 'y' = $M_A - M_C$ = (1046 − 428) × 5.0/4.5 = 686 kNm, so that M_C = 1046 − 686 = 360 kNm. It then follows from Section 2.6.4 that

$$x = \frac{ay}{b} = \frac{686 \times 1.25}{5} = 172 \text{ kNm}$$

and the second part of the reactant line can be drawn from M_C to M_F = 360 − 172 = 188 kNm.

The maximum sagging bending moment between C and F now scales or calculates as 198 kNm. As a conservative judgement was made of the fixing moment above B, and as no significant reduction due to second-order effects is anticipated, a 406 × 140 × 39 UB with M_p = 198 kNm can be chosen for the rafter giving the precise capacity required.

The remaining part of the reactant diagram between A and B can now be drawn without calculation as the tie force 'T' is chosen to bring the bending moment back to zero at A. The completed bending moment diagram is shown shaded in Fig. 7.7.

As with all of these examples, a suitable haunch length to avoid plasticity at the rafter end of the haunch can be scaled off the bending moment diagram.

For the assessment of second-order effects, the frame can be considered as a fixed based frame with a height to eaves of 5 m. The calculation of the elastic critical load then proceeds as follows:

I_c = Second moment of area of column = $2.545 \times 10^8 \text{ mm}^4$
I_r = Second moment of area of rafter = $1.241 \times 10^8 \text{ mm}^4$
P_c = Axial force in column = 125 kN
H = Shear in column = $y/5$ = 137.2 kN
P_r = Axial force in rafter = $137.2 \cos 5.711° + 125 \sin 5.711°$
  = 149 kN

$$R = \frac{I_c L_r}{I_r h} = \frac{2.545 \times 10^8 \times 12.56}{1.241 \times 10^8 \times 5} = 5.152$$

$$\lambda_{cr} = \frac{5E(10 + R)}{\dfrac{5P_r L_r^2}{I_r} + \dfrac{2RP_c h^2}{I_c}}$$

$$= \frac{5 \times 205 \times (10 + 5.152)}{\dfrac{5 \times 149 \times 12\,560^2}{1.241 \times 10^8} + \dfrac{2 \times 5.152 \times 125 \times 5000^2}{2.545 \times 10^8}}$$

$$= 14.5$$

As λ_{cr} is greater than 10, second-order effects can be neglected. An exact elastic critical load analysis with sway prevented at the level of the floor beam confirms this value. It should be carefully noted here that if the exact elastic critical load analysis is repeated with the frame free to sway at the floor level, the elastic critical load drops to 1.33. This result emphasises that the stability of the frame is very sensitive to the stiffness

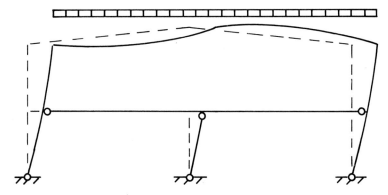

Fig. 7.8 Critical buckling mode of frame with intermediate floor.

of the horizontal restraint provided by the floor. Advantage should only be taken of this restraint if it is completely reliable.

Some readers may find the above conclusion surprising but, although a tie at the intermediate floor level is very effective in preventing the columns from spreading apart at that level and thereby makes a significant contribution to the plastic design capacity under uniform vertical load, it is totally ineffective in resisting the critical buckling mode as shown in Fig. 7.8. A similar situation also arises in the next example.

7.4 Tied portal frame

Economy in portal frame construction is sometimes obtained at the expense of headroom by the provision of a light tie at eaves level. This tie is usually a steel angle or round bar and a typical arrangement is shown in Fig. 7.9.

This arrangement turns the roof into a triangulated structure and induces high axial compressive forces in the rafters. In view of the conclusions reached in previous examples regarding second-order effects, this consideration should sound load alarm bells and it can be anticipated that tied portal frames will have very low elastic critical loads. This indeed proves to be the case and tied portals should *not* be designed by plastic theory. They should rather be designed by elastic theory taking due account of second-order effects. The example which follows is therefore included to demonstrate this conclusion and not as an example to be followed in practice.

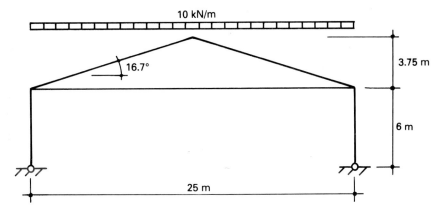

Fig. 7.9 Tied portal.

It is conventional wisdom that the roof slope of a tied portal should not be less than, say, 12° because the shallower the roof slope the larger the axial stresses in both the tie and the rafter, and designs with low roof slopes are impracticable. The stretching of the tie under axial load allows the eaves to spread and the apex to drop. Even with sufficient roof slope, it may be necessary to detail the structure to allow site adjustment so that the stanchions can be made to be near vertical under a predefined portion of the dead load.

In any case, the connection of the tie at the eaves requires a detail that allows the length to be adjusted in order to accommodate the fabrication tolerances, particularly if it is connected to an eaves haunch cutting or if intermediate tie connections have bolts in shear in normal oversized holes.

Under vertical loading, the stanchions do not carry large bending moments and therefore the bases can be either pinned or fixed and this will not affect the design. However, tied portals are very sensitive to transverse or asymmetric loading and fixed bases will significantly improve the stability. For the same reason, the notional horizontal loads are much more likely to influence the design than in conventional portal frame construction. Asymmetric snow loads on roof slopes greater than 15° and exceptional snow drifting in the valleys of multi-span structures also need consideration.

Load combinations involving wind with no imposed load may well put the tie into compression. In such cases, the analysis should be repeated ignoring the tie and treating the structure as an open portal.

7.4.1   Example 7.3. Tied portal

The frame and its loads at the ultimate limit state are shown in Fig. 7.9 and the single relevant collapse mechanism under uniform vertical loading is shown in Fig. 7.10.

Tied portals are usually designed without long haunches and with the same section for the rafters and stanchions. The required full plastic moment M_p is given by

$$125 \times 3.125\theta = 4M_p\theta$$

$$\therefore \quad M_p = 97.66 \, \text{kNm}$$

For illustrative purposes, therefore, we will choose a $305 \times 102 \times 28$ UB with $M_p = 112 \, \text{kNm}$ although this section has a low *lateral* stability and, particularly in the presence of high axial load, would require close purlin centres in order to avoid lateral torsional buckling. Conversely,

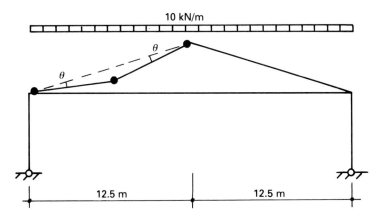

Fig. 7.10 Collapse mechanism.

the relatively deep section is advantageous with regard to the *in-plane* buckling associated with second-order effects.

The forces on one half of the frame are shown in Fig. 7.11 and the unknowns, H, T and P_r, can be determined from considerations of equilibrium.

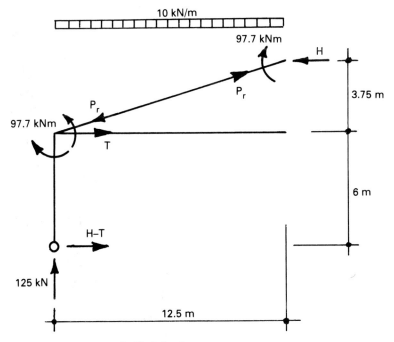

Fig. 7.11 Forces on one half of the frame.

From moment equilibrium of the rafter

$$125 \times 6.25 - 97.7 + 97.7 - 3.75H = 0$$

$$\therefore \quad H = \frac{125 \times 6.25}{3.75} = 208.3\,\text{kN}$$

and $\quad P_r = \dfrac{H}{\cos 16.7°} = 217.5\,\text{kN}$

and from moment equilibrium of the stanchion

$$6(H - T) = 97.7\,\text{kNm}$$

$$\therefore \quad T = 208.3 - \frac{97.7}{6} = 192\,\text{kN}$$

The elastic critical load can now be determined using the same equation as for the portal frame without the tie. This is because the tie is of no benefit as far as the critical buckling mode shown in Fig. 7.12 is concerned. Thus:

h	= Height to eaves	= 6 m
L_r	= Length of rafter	= 13.05 m
I_c	= Second moment of area of column	= $0.5439 \times 10^8\,\text{mm}^4$
I_r	= Second moment of area of rafter	= $0.5439 \times 10^8\,\text{mm}^4$
P_c	= Axial force in column	= 125 kN
P_r	= Axial force in rafter	= 217.5 kN

$$R = \frac{I_c L_r}{I_r h} = \frac{0.5439 \times 10^8 \times 13.05}{0.5439 \times 10^8 \times 6} = 2.175$$

$$\lambda_{cr} = \frac{3EI_r}{I_r \left[\left(1 + \dfrac{1.2}{R} \right) P_c h + 0.3 P_r L_r \right]}$$

$$= \frac{3 \times 205 \times 0.5439 \times 10^8}{13\,050 \left[\left(1 + \dfrac{1.2}{2.175} \right) \times 125 \times 6000 + 0.3 \times 217.5 \times 13\,050 \right]}$$

$$= 1.27$$

This elastic critical load is of course so low that plastic design is out of the question as anticipated in the introduction to this example. Logically, this will always be the case and hence the conclusion that tied portal frames are not suitable for plastic design.

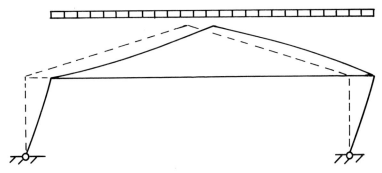

Fig. 7.12 Critical buckling mode of tied portal frame.

7.5   Monitor roof portal frame

The monitor roof portal frame is popular where natural daylight is required but with limited direct sunlight penetration. It is therefore more suited to the tropical zones unless overhangs are incorporated in the roof.

As a consequence of the vertical legs in the monitor roof, these frames have more complicated bending moment diagrams than conventional portal frames. This results in a greater number of potential plastic hinge positions and a greater variety of possible collapse mechanisms. However, these are not difficult to deal with when graphical methods of analysis are used.

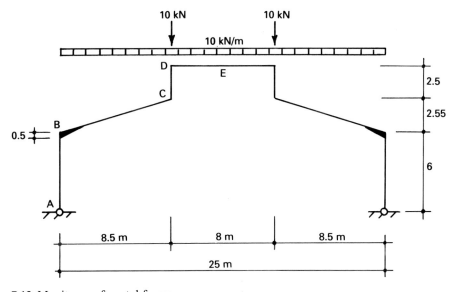

Fig. 7.13 Monitor roof portal frame.

7.5.1 *Example 7.4. Monitor roof portal*

A monitor roof portal frame with its ultimate limit state loads is shown in Fig. 7.13.

Because of the more complex geometry, it is easier to make this type of frame statically determinate by releasing the horizontal force at the base and working with the single redundant H as shown in Section 2.7 and Fig. 2.76. The complete bending moment diagram is given in Fig. 7.14 which is constructed as follows.

As this frame is only carrying vertical forces, the free bending moment diagram is the same as that of a simply supported beam. The vertical reactions at the bases are both 135 kN. As the frame and loading are both symmetrical, it is sufficient to consider half of the frame.

Between A and B, $M_{AB} = 0$

Between B and C, with x measured from B in plan,

$$M_{AB} = 135x - \frac{10x^2}{2}$$

At C, $x = 8.5$ m and this gives a free bending moment of 786.25 kNm which remains constant between C and D.

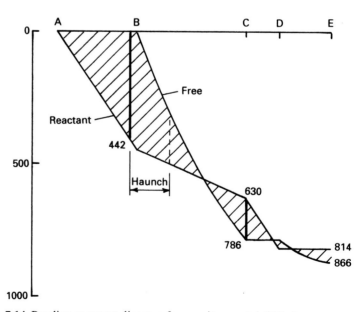

Fig. 7.14 Bending moment diagram for monitor portal (kNm).

Between D and E, with x measured from B in plan,

$$M_{DE} = 135x - \frac{10x^2}{2} - 10(x - 8.5) = 85 + 125x - 5x^2$$

At E, $x = 12.5\,\text{m}$ and $M_E = 866.25\,\text{kNm}$.

The reactant bending moment at any point is equal to the height of that point above the base multiplied by H. The design is initiated by guessing a size for the stanchion and here we choose a $457 \times 191 \times 67\,\text{UB}$ with $M_p = 405\,\text{kNm}$. If we assume a plastic hinge below the haunch, where the height above the base is $5.5\,\text{m}$

$$H = \frac{405}{5.5} = 73.64\,\text{kN}$$

The remainder of the reactant line then follows:

At B	$6H = 441.8\,\text{kNm}$
At C	$8.55H = 629.6\,\text{kNm}$
At D and E	$11.05H = 813.7\,\text{kNm}$

It is easy to see that the design of the rafter is governed by the bending moment at C which equals $786.3 - 629.6 = 156.7\,\text{kNm}$. We can therefore choose a $406 \times 140 \times 39\,\text{UB}$ with $M_p = 198\,\text{kNm}$. The length of the haunch can now be calculated or scaled from the bending moment diagram in order to avoid plasticity at its rafter end.

There are no simple formulae for checking second-order effects in monitor portals and it is necessary to adapt the equations given in Chapter 3 in order to obtain a reasonable estimate. The critical buckling mode is again a combined stanchion and rafter mode but the situation is complicated by the geometric effect of the monitor which inhibits the free buckling of the rafter and also dissipates some of the axial load in this member. If the rafter is 'unwrapped', its total length is $L_r = 8.87 + 2.5 + 4.0 = 15.37\,\text{m}$. However, to include this length with the full axial force P_r in the rafter, calculated at the eaves, would be excessively conservative. It is sufficient to consider L_r to comprise the sloping length of the rafter together with the vertical leg of the monitor and to combine this with the calculated value of P_r. The calculation then proceeds as follows.

h	= Height to eaves	$= 6\,\text{m}$
L_r	= Effective length of rafter	$= 8.87 + 2.5 = 11.37\,\text{m}$
I_c	= Second moment of area of column	$= 2.941 \times 10^8\,\text{mm}^4$
I_r	= Second moment of area of rafter	$= 1.241 \times 10^8\,\text{mm}^4$

$$P_c = \text{Axial force in column} \qquad = 135 \text{ kN}$$
$$H = \text{Base shear} \qquad = 73.64 \text{ kN}$$
$$P_r = \text{Axial force in rafter} \qquad = H\cos 16.7° + P_c \sin 16.7°$$
$$= 109.3 \text{ kN}$$

$$R = \frac{I_c L_r}{I_r h} = \frac{2.941 \times 10^8 \times 11.37}{1.241 \times 10^8 \times 6} = 4.491$$

$$\lambda_{cr} = \frac{3EI_r}{L_r \left[\left(1 + \frac{1.2}{R}\right) P_c h + 0.3 P_r L_r \right]}$$

$$= \frac{3 \times 205 \times 1.241 \times 10^8}{11\,370 \left[\left(1 + \frac{1.2}{4.491}\right) \times 135 \times 6000 + 0.3 \times 109.3 \times 11\,370 \right]}$$

$$= 4.80$$

The value calculated in this way is conservative with respect to the exact value of $\lambda_{cr} = 5.30$.

It follows that second-order effects can be treated approximately and that the minimum value of the load factor λ_p at plastic collapse is given by

$$\lambda_p = \frac{0.9\lambda_{cr}}{\lambda_{cr} - 1} = \frac{0.9 \times 4.80}{3.80} = 1.137$$

As the actual value of λ_p provided by the proposed design is about 1.05, a stronger and stiffer rafter should be chosen and the stability check repeated.

This example adds further evidence that irregular portal frames should be treated with respect as far as second-order effects are concerned. Prior to the study reported in this chapter, there had been no indication of just how sensitive to in-plane buckling these structures can be.

7.5.2   Example 7.5. Monitor portal with exceptional snow load

The monitor roof portal, with its abrupt change of roof levels, will have an exceptional snow load case to be considered. Figure 7.15 shows the frame of Example 7.4 with the loads assumed for this load case at the ultimate limit state.

The frame is once more made statically determinate by releasing the horizontal reaction at the right-hand base and the free bending moment

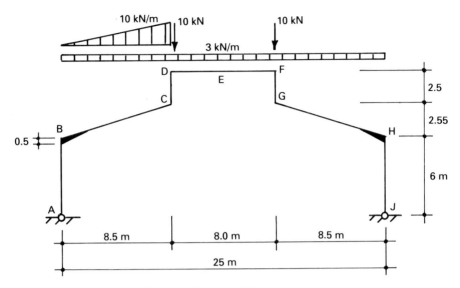

Fig. 7.15 Exceptional snow load on monitor portal frame.

diagram for the roof members between B and H is again that of a simply supported beam of 25 m span subject to the same vertical loads as the roof. The vertical reactions V_A and V_J at A and J can be determined by considering equilibrium:

$$25V_J = 10(8.5 + 16.5) + 75 \times 12.5 + 42.5 \times 5.667$$

$$\therefore \quad V_J = \frac{1428.3}{25} = 57.13 \, \text{kN}$$

and $V_A = 137.5 - 57.13 = 80.37 \, \text{kN}$

The free bending moments are therefore as follows:

Between A and B, $M_{AB} = 0$
Between B and C with x measured from B in plan

$$M_{BC} = 80.37x - \frac{3x^2}{2} - \frac{10x}{8.5} \frac{x}{2} \frac{x}{3} = 80.37x - 1.5x^2 - 0.196x^3 \, \text{kNm}$$

When $x = 8.5$, $M_C = 454.3 \, \text{kNm}$.
Between C and D, M_{CD} is constant.
Between D and E, with x measured from B in plan as above

$$M_x = 80.37x - 42.5(x - 5.67) - 10(x - 8.5) - 1.5x^2$$

$$= 325.8 + 27.87x - 1.5x^2 \, \text{kNm}$$

so that at E $x = 12.5 \, \text{m}$ and $M_E = 439.8 \, \text{kNm}$

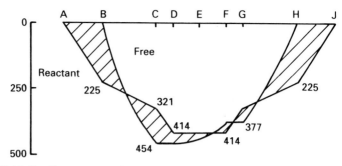

Fig. 7.16 Bending moment diagram for monitor with exceptional snow (kNm).

and at F $x = 16.5\,$m and $M_F = 377.2\,$kNm.
Between J and H, $M_{JH} = 0$
Between H and G, with x measured from H in plan

$$M_{HG} = 57.13x - \frac{3x^2}{2}$$

so that at G, with $x = 8.5\,$m, $M_G = 377.2\,$ kNm.
Finally, between G and F, M_{GF} is constant $= 377.2$ kNm and the calculations for the two halves of the frame agree.

The free bending moment diagram can now be drawn as shown in Fig. 7.16.

The reactant line can now be drawn for any suitable value of the base shear H and the one shown in Fig. 7.16 corresponds to $H = 37.5\,$kN giving

$$
\begin{aligned}
M_B &= 37.5 \times 6 &&= 225.0\,\text{kNm} \\
M_C &= 37.5 \times 8.55 &&= 320.6\,\text{kNm} \\
M_D = M_E &= 37.5 \times 11.05 &&= 414.4\,\text{kNm}
\end{aligned}
$$

In comparison with Fig. 7.14, the bending moments in Fig. 7.16 are quite small and it is evident that the exceptional snow load case is not critical. This case need not, therefore, be pursued further.

7.6 North light portal frames

A north light portal frame is a non-symmetrical portal frame, a typical example being shown in Fig. 7.17. The asymmetry does not introduce any unexpected complexities except that the sag of the long rafter should be checked.

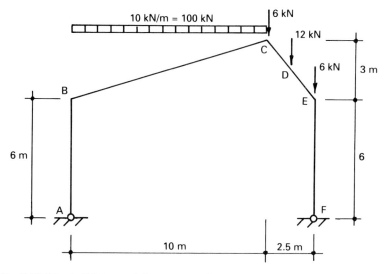

Fig. 7.17 North light portal frame example.

7.6.1   Example 7.6. North light portal frame

The frame is shown in Fig. 7.17 together with its loading at the ultimate limit state.

Because the frame is not symmetrical, it is again probably easier to make the frame statically determinate by releasing the right-hand base rather than making a cut at the apex. The vertical reactions at the base are easily determined by equilibrium:

$$12.5V_R = 100 \times 5 + 6 \times 10 + 12 \times 11.25 + 6 \times 12.5$$

$$\therefore \quad V_R = \frac{770}{12.5} = 61.60 \, \text{kN}$$

and $\quad V_L = 124 - 61.60 = 62.40 \, \text{kN}$

The free bending moments then follow in much the same way as the previous example:

Between A and B, $M_{AB} = 0$
Between B and C, with x measured from B in plan

$$M_{BC} = 62.40x - \frac{10x^2}{2}$$

At C, where $x = 10 \, \text{m}$, $M_C = 624 - 500 = 124 \, \text{kNm}$.
Between F and E, $M_{FE} = 0$
$M_D = 1.25(61.60 - 6) = 69.5 \, \text{kNm}$
$M_C = 2.5(61.6 - 6) - 1.25 \times 12 = 124 \, \text{kNm}$ as before.

The free bending moment diagram can now be drawn as shown in Fig. 7.18.

This is a frame of relatively small span so that a uniform frame with only nominal haunches is appropriate. Once more, we commence the design by assuming a plastic hinge at the eaves (either B or E) and we try a $305 \times 102 \times 25$ UB with a full plastic moment of $92.4\,\text{kNm}$. This gives

$$H = \frac{92.4}{6} = 15.4\,\text{kN}$$

and, for the reactant bending moment diagram, which is also drawn on Fig. 7.18,

$$M_D = 15.4 \times 7.5 = 115.5\,\text{kNm}$$
$$M_C = 15.4 \times 9 \quad = 138.6\,\text{kNm}$$

The maximum bending moment in the rafter scales about $74\,\text{kNm}$ so that the plastic design of the uniform frame is satisfactory. In fact, the plastic collapse load factor is about 1.09.

If required, the maximum bending moment in the rafter can be calculated exactly. If x is measured in plan from B,

$$M_{BC} = 62.4x - \frac{10x^2}{2} - 15.4\left(6 + \frac{3x}{10}\right)$$

$$= -92.4 + 57.78x - 5x^2 \,\text{kNm}$$

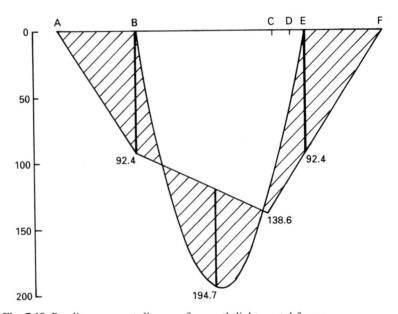

Fig. 7.18 Bending moment diagram for north light portal frame.

The maximum bending moment is when

$$\frac{\mathrm{d}M}{\mathrm{d}x} = 0 \qquad \text{i.e.} \quad x = 5.778\,\mathrm{m}$$

which gives $M_{\max} = -92.4 + 333.9 - 166.9 = 74.5\,\mathrm{kNm}$

Once again, it is necessary to consider second-order effects before the choice of member sizes can be considered to be complete and, once again, the methods given in Chapter 3 are not directly applicable and can be considered to give approximate guidance only. The most reasonable procedure here appears to be to apply the model shown in Fig. 3.11, and the associated equation given in Section 3.2.3, to the left-hand leg and the long rafter of the frame. As the right-hand rafter is relatively stocky, the right-hand half of the frame will be stiffer than the part considered and will offer it some additional restraint. This procedure is therefore likely to be conservative:

$$
\begin{aligned}
h &= \text{Height to eaves} & &= 6\,\mathrm{m} \\
L_r &= \text{Effective length of rafter} & &= 10.44\,\mathrm{m} \\
I_c &= \text{Second moment of area of column} & &= 0.4364 \times 10^8\,\mathrm{mm}^4 \\
I_r &= \text{Second moment of area of rafter} & &= 0.4364 \times 10^8\,\mathrm{mm}^4 \\
P_c &= \text{Axial force in column} & &= 62.4\,\mathrm{kN} \\
H &= \text{Base shear} & &= 15.4\,\mathrm{kN} \\
P_r &= \text{Axial force in rafter} & &= H\cos 16.7^\circ + P_c\sin 16.7^\circ \\
& & &= 32.68\,\mathrm{kN}
\end{aligned}
$$

$$R = \frac{I_c L_r}{I_r h} = \frac{0.4364 \times 10^8 \times 10.44}{0.4364 \times 10^8 \times 6} = 1.74$$

$$\lambda_{\mathrm{cr}} = \frac{3EI_r}{L_r\left[\left(1 + \dfrac{1.2}{R}\right)P_c h + 0.3 P_r L_r\right]}$$

$$= \frac{3 \times 205 \times 0.4364 \times 10^8}{10\,440\left[\left(1 + \dfrac{1.2}{1.74}\right) \times 62.4 \times 6000 + 0.3 \times 32.68 \times 10\,440\right]}$$

$$= 3.50$$

This value is of course less than the limiting value of 4.6 (see Section 3.2.6) and on this basis the approximate treatment is invalid and second-order effects should be treated more precisely. However, the value calculated in this way is conservative and the more precise value of λ_{cr} calculated using exact analysis is 4.92. Nevertheless, even this higher value of λ_{cr} requires a plastic collapse load factor given by

$$\lambda_p = \frac{0.9\lambda_{cr}}{\lambda_{cr} - 1} = \frac{0.9 \times 4.92}{3.92} = 1.13$$

which is higher than the 1.09 provided. It follows that a uniform frame of $305 \times 102 \times 25$ UB is not quite adequate and some strengthening is required.

This example emphasises once again that irregular frames tend to be sensitive with regard to second-order effects and that these should always be carefully considered. The intelligent use of the approximate methods given in Chapter 3 will usually give an indication of the seriousness of the problem and will often also offer a sufficiently accurate solution to enable a safe design to be obtained.

Thus, here, an increase in member size will both strengthen and stiffen the frame and make it possible to satisfy the conservative approximate treatment of second-order effects used above. Alternatively, fixed bases would reduce the slenderness of the relatively long columns and be very beneficial in increasing the elastic critical load to an acceptable level.

7.7   Lean-to portal frames

A lean-to portal structure is often built alongside or attached to an existing structure. A typical shape is shown in Fig. 7.19. The design procedures and other practical considerations are similar to those for the north light portal considered in the previous section.

An upper eaves haunch can be beneficial in reducing the rafter sag. If the upper eaves is rigidly connected to a stronger, stiffer frame, it can also reduce the rafter size. However, in such a case, the bending moments arising from this attachment can be sufficiently large to cause a plastic hinge in the upper eaves zone early in the loading history.

In the example shown in Fig. 7.19, which will be designed later, the long slender right-hand column indicates possible problems with second-order effects. Base fixity would be of considerable benefit here and would also produce an overall increase in stiffness and strength.

7.7.1   *Example 7.7. Lean-to portal frame*

The frame for this example is shown in Fig. 7.19. It is a stand-alone frame but the considerations would be similar if it were connected at the right-hand eaves to an existing structure.

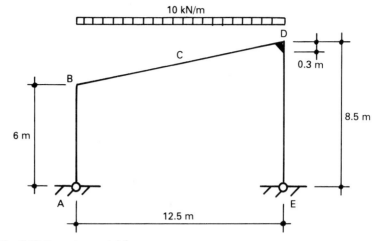

Fig. 7.19 Lean-to portal frame.

The procedure for plastic design is very similar to that used in the previous example of a north light portal, leading to the bending moment diagram shown in Fig. 7.20. When the right-hand base is released, the equation for the free bending moment in the rafter is

$$M_{BD} = \frac{10x}{2}(12.5 - x)$$

If we assume a uniform frame with a plastic hinge below the small haunch at D, and guess a $305 \times 102 \times 28$ UB with $M_p = 112\,\text{kNm}$, the base shear H is given by

$$H = \frac{112}{8.2} = 13.66\,\text{kN}$$

The reactant moment line is then drawn from

$$M_B = 13.66 \times 6 \quad = 81.95\,\text{kNm}$$
$$M_D = 13.66 \times 8.5 = 116.1\,\text{kNm}$$

and the resulting maximum bending moment in the rafter scales about 96 kNm. The plastic collapse load factor given by this design is 1.08.

A suitable approximate treatment for second-order effects is to assume a point of contraflexure in the middle of the rafter and to use the equations of Section 3.2.3. A moment's reflection is sufficient to make it clear that the columns will have much more influence than the rafter and that it is sufficient to consider only the right-hand half of the frame.

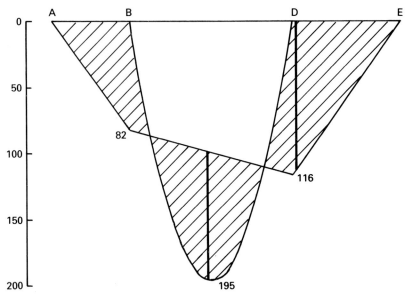

Fig. 7.20 Bending moment diagram for lean-to portal frame (kNm).

This is likely to prove conservative as the stiffer left-hand half of the frame will provide additional restraint:

h = Height to eaves = 8.5 m

L_r = Length of rafter to mid-span = 6.37 m

I_c = Second moment of area of column = 0.5439×10^8 mm^4

I_r = Second moment of area of rafter = 0.5439×10^8 mm^4

P_c = Axial force in column = 62.5 kN

H = Base shear = 13.66 kN

P_r = *Average* axial force in rafter = $H \cos 11.3°$

= 13.4 kN

$$R = \frac{I_c L_r}{I_r h} = \frac{0.5439 \times 10^8 \times 6.37}{0.5439 \times 10^8 \times 8.5} = 0.75$$

$$\lambda_{cr} = \frac{3EI_r}{L_r\left[\left(1 + \dfrac{1.2}{R}\right)P_c h + 0.3 P_r L_r\right]}$$

$$= \frac{3 \times 205 \times 0.5439 \times 10^8}{6370\left[\left(1 + \dfrac{1.2}{0.75}\right) \times 62.5 \times 8500 + 0.3 \times 13.4 \times 6370\right]}$$

$$= 3.73$$

This value is again less than the limiting value of 4.6 (see Section 3.2.6) and, on this basis, second-order effects require an increase in the member sizes. The exact elastic critical load is 5.37, which is a considerable improvement on the approximate value calculated above. However, it is difficult to suggest a better approximate treatment than the one used. Even the higher elastic critical load value would, strictly speaking, still require a redesign because the necessary value of the load factor is

$$\lambda_p = \frac{0.9\lambda_{cr}}{\lambda_{cr} - 1} = \frac{0.9 \times 5.37}{4.37} = 1.11$$

which is higher than the 1.08 provided.

Reference

7.1 Steel Construction Institute. *Steelwork Design*: Guide to BS 5950: Part 1: 1990, Volume 1 – Section properties and member capacities, 3rd ed. 1992.

Chapter 8
Multi-span Portal Frames

8.1 General

Multi-span portal frames are a common form of structure and can include a number of spans ranging from two to ten or more. The design effort and the level of rigour required to design structures of this type means that computer programs are now very desirable if not absolutely necessary. In particular, the increased complexity means that the plastic hinge history cannot be determined manually and the designer requires this information in order to

❑ allow the member stability checks to be carried out at the minimum required ultimate limit state design loading rather than at the higher collapse mechanism loading;
❑ check for any transient plastic hinges and to determine whether or not they should be considered in the member stability calculations;
❑ determine if any plastic hinges have formed at the serviceability limit state loadings and whether these should be taken into account in the deflection calculations.

However, blind faith in computer programs is a recipe for disaster and it is considered essential (Reference 8.1) to be able to carry out sufficient manual checks to verify the computer solution. This is only possible if the designer has sufficient understanding of the principles involved to be able to choose and analyse a suitable 'checking model'. It follows that even if the designer has available a computer program for elastic-plastic analysis, it may still be necessary to make reference to this chapter in order to carry out the necessary checks.

Virtually all multi-span portal frames have valleys and therefore the design will include the exceptional snow loading case. Furthermore, frames with roof slopes greater than 15° will also have the asymmetric snow loading case. Hence there is the possibility of three separate snow loading cases.

The natural behaviour of portal frames gives rise to eaves spread under vertical load. The presence of multiple bays tends to aggravate this tendency and it is necessary to give some consideration to the steel erection and roof cladding erection sequences. Steelwork erection sequences can sometimes require internal span members to act as temporary side span members and, if these members have less stiffness, then additional deflections may well occur during erection. These deflections can be increased by the early erection of the roof cladding and can then be locked in by the roof cladding acting as a stressed skin diaphragm. Frames with steeper roof slopes have a greater potential for eaves spread.

Precambering the frames by changing the member connection angles and shortening the rafter lengths can help produce better verticality in the outer stanchions at the dead loading stage. Tight tolerances for portal frame side stanchion head positions should not be specified because small eaves movements occur all the time due to wind uplift loads and temperature changes as well as snow loading.

The designer should also consider lateral deflections at the internal stanchion heads, particularly when some frames are on valley beams. In practice, transverse wind loads should not cause any problems because the diaphragm action of the roof cladding is particularly effective in this situation. However, it is difficult and time-consuming to prove roof diaphragm action and most designers prefer not to have to carry out this calculation. Those who do should refer to BS 5950: Part 9[8.2]. In-plane roof bracing along some valley lines is a popular alternative method of minimising differential transverse deflections and also serves as an erection aid. As with diaphragm action, roof bracing also reduces the unfavourable influence of second-order effects.

Notwithstanding the difficulties, this chapter describes the manual analysis of multi-span portal frames and the following principles will help the designer to understand the examples which follow.

❑ Unless the internal rafters are unreasonably small, the collapse mechanism under uniform vertical load will be an outer bay mechanism as illustrated in Fig. 8.1. Depending on the length of the haunch, the outer eaves hinge may form either in the stanchion below the haunch or at the rafter end of the haunch. However, the hinge associated with the internal haunch will *always* form at the rafter end of the haunch, regardless of its length. It follows that, in contrast to single-bay frames, with multi-bay frames it is impossible to avoid having plastic hinges in the rafters at the end of the haunches however undesirable this may be considered to be.

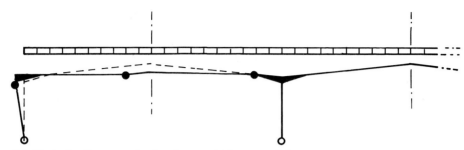

Fig. 8.1 Typical collapse mechanism in a multi-bay portal frame.

❑ In general, an internal bay collapse mechanism will only form *after* the prior formation of an outer bay mechanism such as the one shown in Fig. 8.1. Any exception to this rule would require the formation of a rafter mechanism as shown for a tied portal in Fig. 7.10. However, as discussed in Section 7.3, portal frames which fail with this type of mechanism usually have to be rejected because of second-order effects. The 'snap-through' check in clause 5.5.3.3 of BS 5950: Part 1 is therefore important and should not be over-looked.

❑ For design, a multi-bay frame is generally best made statically determinate by a series of apex cuts as shown in Fig. 8.2. The geometric conditions derived in Section 2.6.4 for the reactant bending moments in a single-bay frame can then be applied independently to each bay in turn and the internal stanchions will be subjected to reactant bending moments from the two adjacent bays superimposed upon each other. These often tend to cancel each other out with the result that the internal stanchions may be relatively lightly stressed.

❑ There are a variety of ways of drawing the reactant bending moment diagram corresponding to the redundant forces shown in Fig. 8.2. The examples which follow illustrate some of the available procedures. There are two particular points to note:

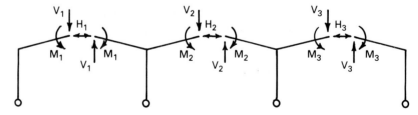

Fig. 8.2 Reactant forces in a multi-bay frame.

(1) It is essential to ensure moment equilibrium in the internal stanchions. In a pinned based structure, this means that *either* the base moment is zero in each individual bay *or* non-zero base moments in adjacent bays are designed to be in balance.

(2) If the reactant line can be made to be symmetrical in a particular bay, $V = 0$ and it is much easier to satisfy the requirements of equilibrium.

If there are no valley stanchions in a frame so that the valleys are supported on valley beams, then the horizontal redundancy H will be common across a series of spans until a stanchion or roof plane bracing occurs. Also the eaves bending moments adjacent to valley beams must balance each other because valley beams cannot support torsional bending moments. These considerations provide additional constraints to the reactant bending moment diagram.

Frames supported on valley beams which have valley roof plane bracing should ideally be analysed as a three-dimensional structure. With a little ingenuity, this can be done using an elastic or elastic-plastic plane frame analysis program by arranging the frames side by side so that they occupy the same plane and by linking them at the bracing lines. This procedure has been explained in the context of stressed skin design[8.3] and precisely the same principles apply here.

If a computer program is not used to provide an elastic-plastic analysis, including determination of the hinge history, then it is recommended that no plastic hinges should be permitted to form at the serviceability limit state loadings. This can be verified by a linear-elastic analysis. In this case, the frame has to be checked for member stability at the collapse mechanism loading rather than at the minimum required design loading.

8.2   Example 8.1. Three-span portal frame

The general arrangement of this example is shown in Fig. 8.3(a) with its required ultimate limit state loading for the symmetrical snow case. The exceptional snow loading case is shown in Fig. 8.3(b).

The symmetrical snow loading case is considered first.

8.2.1   *Design for symmetrical snow loading*

The frame is first made statically determinate by three apex cuts as shown in Fig. 8.2. The construction of the free bending moment diagram

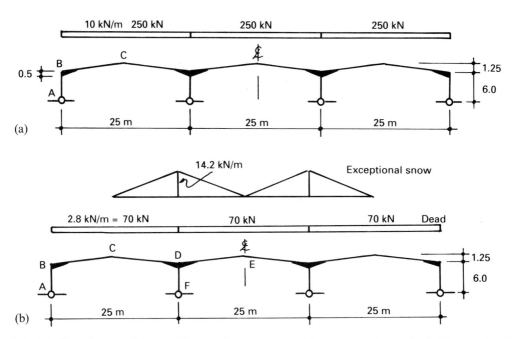

Fig. 8.3 Three-bay portal frame: (a) general arrangement with symmetrical snow load; (b) exceptional snow load case.

is then very simple and is unchanged from earlier single-bay examples. Because of symmetry, it is sufficient to consider half of the structure and the resulting bending moment diagram is shown in Fig. 8.4 with a free bending moment of $250 \times 25/8 = 781.25$ kNm at each eave and valley. There is no free bending moment in the internal stanchions.

When the spans are identical it is worth considering making the rafter sections and eaves haunches identical in order to save on design office, drawing office and fabrication time unless the fabricator has sophisticated computer facilities that can accommodate the required variations. However, as the valley stanchions are almost certainly smaller than the side stanchions, then the haunch lengths on the bending moment diagram will not be identical.

The construction of the reactant bending moment line in Fig. 8.4 can be a little tricky and there is more than one way to do it. We bear in mind that the side spans and centre span can have different apex redundancies as long as the resulting bending moments are in balance at the pinned bases of the internal stanchions. One possible method is first to choose a size for the outer stanchion with a plastic hinge just below the haunch as was done for a number of the examples considered in the previous chapter. This fixes point 'd' in Fig. 8.4.

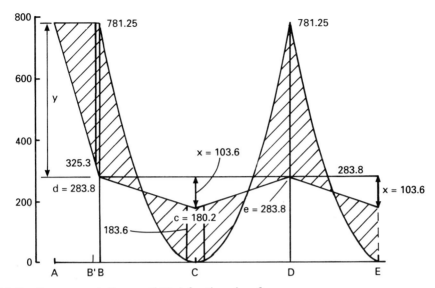

Fig. 8.4 Bending moment diagram (kNm) for three-bay frame.

Here, we select a $457 \times 191 \times 74$ UB with a full plastic moment of 456 kNm. Ordinate 'd' is then

$$781.25 - 456 \times \frac{6.0}{5.5} = 781.25 - 497.45 = 283.8 \text{ kNm}$$

This also fixes dimension 'x' (see also Fig. 2.72) as

$$x = \frac{ay}{b} = \frac{1.25 \times 497.45}{6.0} = 103.6 \text{ kNm}$$

At this stage, only one further point is necessary in order to completely define the reactant bending moment diagram for the outer rafter. The easiest solution by far is to simply assume the bending moment diagram to be symmetrical as far as the outer bay is concerned as shown in Fig. 8.4. This is equivalent to assuming the vertical force at the apex cut (V_1 in Fig. 8.2) to be zero. However, as we shall see later, this may not necessarily be the best solution though we will continue with it for the time being.

This assumption gives the ordinate of point 'c' in Fig. 8.4 as $283.8 - 103.6 = 180.2$ kNm. Some designers may prefer to work with graphical constructions in the manner illustrated above. However, for the majority of the examples in this chapter, it is easier to work with the redundant forces M, H and V as unknowns in the equilibrium equations.

Noting that $V_1 = 0$ and that the calculation is carried out with the convention that moments causing tension on the outside of the frame are positive, the above calculation then becomes:

At A $\quad 0 \quad = -M_1 \quad -7.25H_1 \quad +781.25$

At B' $\quad 456 = -M_1 \quad -1.75H_1 \quad +781.25$

giving: $\quad H_1 = 82.9\,\text{kN}$

$\quad\quad\quad M_1 = 180.2\,\text{kNm}$

The equation for the bending moment in rafter is now

$$M_r = -M_1 - 0.1 \times H_1 + 5x^2 = -180.2 - 8.29x + 5x^2$$

so that the maximum bending moment is where

$$\frac{\mathrm{d}M_r}{\mathrm{d}x} = 0 \quad \text{i.e. when } -8.29 + 10x = 0 \quad \text{or } x = 0.829\,\text{m}$$

and the maximum bending moment in the rafter is

$$M_r^{\max} = -180.2 - 8.29 \times 0.829 + 5 \times 0.829^2 = -183.6\,\text{kNm}$$

Here, we could choose for the rafter section a $406 \times 140 \times 39$ UB with a full plastic moment of 198 kNm but this would undoubtedly lead to problems when second-order effects are considered. A more suitable choice of rafter is therefore a $406 \times 140 \times 46$ UB with a full plastic moment of 245 kNm. Conventional wisdom would now suggest that the bending moment at the end of the haunches should be restricted to $0.87 \times 245 = 213.2\,\text{kNm}$ so that the haunch length could either be scaled off Fig. 8.4 or obtained as the solution of

$$-180.2 - 8.29x + 5x^2 = 213.2\,\text{kNm}$$

which gives $x = 9.74\,\text{m}$ and makes for rather long haunches with a length of $12.5 - x = 2.76\,\text{m}$.

However, conventional wisdom would be wrong because, as explained in connection with Fig. 8.1, regardless of the length of the haunch, it is impossible to avoid a plastic hinge at the end of the valley haunch and there is no point whatsoever in sizing this haunch to maintain an elastic bending moment in the bending moment distribution of Fig. 8.4. This is a statically admissible bending moment distribution but, as drawn, it does not include a valid mechanism and, as explained earlier, the only valid mechanisms all include this plastic hinge.

At the very least, the length of the valley haunch could be reduced to that given by the solution of the following equation with no disadvantage whatsoever.

$$-180.2 - 8.29x + 5x^2 = 245\,\text{kNm}$$

This reduces the required length of the haunch to 2.41 m with a useful saving in fabrication costs.

To complete this solution, it is necessary to consider also the bending moment distribution in the internal bay. This is trivial here as there is no reason to make this distribution any different from that already obtained for the end-bay and Fig. 8.4 has been drawn on the assumption that the internal rafter has the same bending moment distribution as the outer rafter. This assumption is statically admissible and has the advantage that the bending moments in the internal stanchions balance exactly to zero so that there is no need to give further consideration to the equilibrium of these members.

The choice of a suitable member for the internal stanchions is then a matter for some judgement. The only primary design requirement is that these members should be stable under the relatively modest axial loads that they carry. However, a simple prop is not recommended and, indeed, designers are advised not to make the internal stanchions too small as this could aggravate second-order effects and the performance under asymmetrical loading. The internal stanchion head detail will, in many cases, determine the size of this stanchion. The flange and the web must be adequate for the rafter haunch tensile bolt loads and other local loads. Valley beams meeting at the same level as the rafter haunches may require an even deeper stanchion section in order to allow practical connection details, especially when the valley beams also require moment connections.

It then follows that the design of all of the rafters is identical though there will be some detailing differences because the size of the internal stanchions will be less than that of the outer stanchions. Unless an elastic-plastic analysis is carried out in order to determine the hinge history, in addition to providing restraint to the plastic hinges in the outer stanchions below the haunches, it is also necessary to stabilise possible plastic hinges at the rafter ends of all of the haunches and near the apex in all of the spans. Taken together with the rather long haunches, this is unlikely to constitute an economic design.

An alternative approach, which is likely to lead to a more favourable design, is to accept that there will be plastic hinges at the rafter ends of the haunches and to start by assuming a more suitable haunch length together with plastic hinges at these points. Let us therefore consider the effect of starting by assuming the same rafter, namely a $406 \times 140 \times 46$ UB with a full plastic moment of 245 kNm, and a haunch length of 2 m. The bending moment diagram for this case is shown in Fig. 8.5. This procedure too has the advantage that the same bending moment diagram is used for both the internal and external bays

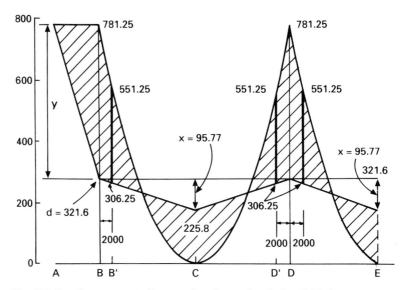

Fig. 8.5 Bending moment diagram for alternative design (kNm).

of the structure. The bending moments in the internal stanchions are in balance and the same design is obtained for any number of spans greater than or equal to two.

The free bending moment part of the diagram is the same as before and the fixed points for the construction of the reactant moment line are the moments of $551.25 - 245 = 306.25$ kNm at the ends of the haunches. The reactant line could be drawn by trial and error or by a formal geometric construction but is best calculated using the relevant equilibrium equations noting that $V_1 = 0$ as before:

$$
\begin{aligned}
\text{At A}\quad & 0 \;= -M_1 \quad -7.25H_1 \quad +781.25\\
\text{At B}'\quad & 245 = -M \quad -1.05H_1 \quad +551.25\\
\text{giving:}\quad & H_1 \;= 76.6\,\text{kN}\\
& M_1 = 225.8\,\text{kNm}\\
\text{and}\quad & M_B = -M_1 - 1.25H_1 + 781.25\\
& \quad\;\; = -225.8 - 1.25 \times 76.6 + 781.25 = 459.7\,\text{kNm}
\end{aligned}
$$

The equation for the bending moment in the rafter is:

$$ M_r = -M_1 - 0.1 \times H_1 + 5x^2 = -225.8 - 7.66x + 5x^2 $$

which has a maximum value of 228.7 kNm at $x = 0.766$ m.

The design moment in the stanchion below the haunch is $459.7 \times 5.5/6.0 = 421.4$ kNm which allows the same choice of stanchion as before, namely a $457 \times 191 \times 74$ UB with a full plastic moment of 456 kNm. However, this moment is greater than the yield moment of the member

(402 kNm) so that it will be necessary to give some consideration to the torsional restraints required to ensure the lateral torsional stability of the partially plastic member. This design process therefore leads to the same choice of members as the first but with shorter haunch lengths.

Another alternative solution is obtained if the designer is prepared to wrestle with the additional complexities of a non-symmetrical bending moment diagram for the outer bay ($V_1 \neq 0$ in Fig. 8.2). This is included here partly to illustrate these additional considerations for the benefit of the interested reader.

We make the same start as in the first solution by assuming a $457 \times 191 \times 74$ UB stanchion with a plastic hinge below the haunch. This fixes ordinate 'd' as 283.8 kNm and 'x' as 103.6 kNm as before and as shown in Fig.8.6. Again, only one further point is necessary in order to define the reactant bending moment diagram for the outer rafter. As a plastic hinge at the rafter end of the internal haunch is unavoidable, this point is chosen. If we use the same rafter section, namely a $406 \times 140 \times 46$ UB with a full plastic moment of 245 kNm and keep the reduced length of the valley haunch at 2.0 m, the ordinate on the reactant bending moment diagram of the point corresponding to this haunch length is $551.25 - 245 = 306.25$ kNm.

The reactant line can now be drawn through 'd' and this ordinate while following the geometric constraint $x/y = a/b$ obtained in Section 2.6.4. This requires a small amount of trial and error or iteration or, alternatively, the reactant line can be obtained exactly by simple

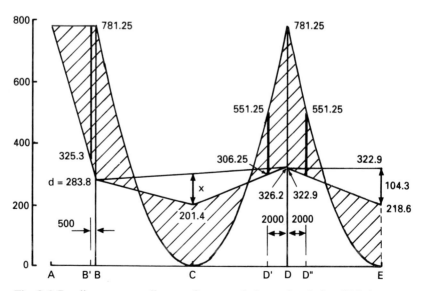

Fig. 8.6 Bending moment diagram for second alternative design (kNm).

geometry. However, once again, the best method appears to be based on solving the equilibrium equations:

$$
\begin{array}{llllll}
\text{At A} & 0 & = -M_1 & -7.25H_1 & +12.5V_1 & +781.25 \\
\text{At B'} & 456 = -M_1 & -1.75H_1 & +12.5V_1 & +781.25 \\
\text{At D'} & 245 = -M_1 & -1.05H_1 & -10.5V_1 & +551.25
\end{array}
$$

giving: $H_1 = 82.9\,\text{kN}$
$V_1 = 1.70\,\text{kN}$
$M_1 = 201.4\,\text{kNm}$

and
$$
\begin{aligned}
M_B &= -M_1 - 1.25H_1 + 12.5V_1 + 781.25 \\
&= -201.4 - 1.25 \times 82.9 + 12.5 \times 1.70 + 781.25 \\
&= 497.5\,\text{kNm} \\
M_D &= -M_1 - 1.25H_1 - 12.5V_1 + 781.25 \\
&= -201.4 - 1.25 \times 82.9 - 12.5 \times 1.70 + 781.25 \\
&= 455.0\,\text{kNm}
\end{aligned}
$$

The maximum sagging bending moment in the right-hand rafter can now be calculated or scaled as about 206 kNm so that the 245 kNm provided by the rafter section selected previously is still sufficient. In the left-hand rafter, the equation for the bending moment is:

$$M_r = -201.4 - (0.1 \times 82.9 + 1.70)x + 5x^2$$

and equating this to $0.87 \times 245 = 213.2\,\text{kNm}$ gives the required haunch length at the outer eaves as 2.71 m.

It is now necessary to address the requirements for the internal rafter and, in order to do this, we need also to consider the equilibrium of bending moments in the internal stanchion.

In the left-hand bay, the bending moment at F is

$$M_F = -M_1 - 7.25H_1 - 12.5V_1 + 781.25$$

$$= -201.4 - 7.25 \times 82.9 - 12.5 \times 1.70 + 781.25 = -42.4\,\text{kNm}$$

For equilibrium, it is necessary that the reactant bending moments from the centre bay neutralise the base moment of 42.4 kNm resulting from the design of the outer bay. The requirements for equilibrium at F, including some values yet to be calculated, are summarised in Fig. 8.7.

For the centre bay, our point of reference is once more the bending moment at the end of the haunch which is again assumed to have a length of 2.0 m. If, for the time being, we assume the same rafter section as before, the ordinate of the reactant line at the haunch is again 306.25 kNm as shown in Fig. 8.6. The equations necessary to draw the symmetrical reactant line shown for the internal bay are

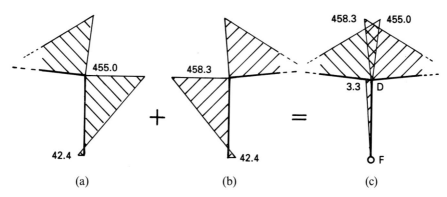

Fig. 8.7 Bending moments in internal stanchion (kNm): (a) from outer bay; (b) from internal bay; (c) resultant moments.

$$
\begin{aligned}
\text{At F} \quad -42.4 &= -M_2 \;\; -7.25H_2 \;\; +781.25 \\
\text{At D''} \quad 245 &= -M_2 \;\; -1.05H_2 \;\; +551.25 \\
\text{giving:} \quad H_2 &= 83.5\,\text{kN} \\
M_2 &= 218.6\,\text{kNm} \\
\text{and} \quad M_D &= -M_1 - 1.25H_1 + 781.25 \\
&= -218.6 - 1.25 \times 83.5 + 781.25 = 458.3\,\text{kNm}
\end{aligned}
$$

The bending moment diagram for the centre bay shown in Fig. 8.6 is thus completed and the maximum bending moment in the rafter is found to be about 222 kNm. The rafter section chosen for the outer bay can therefore again be continued for the internal bay but this time with a considerably reduced haunch length.

It is important to note here that, although we have not obtained a collapse mechanism in the internal bay, the design of the internal and external bays are unavoidably linked through the requirement for equilibrium of the common stanchion. In principle, the strength of the internal rafter could be reduced, taking advantage of the arching effect, until a rafter mechanism of the type illustrated for a tied portal in Fig. 7.10 is obtained. However, arching is only obtained if the necessary thrust is available and this must come from the outer frame. Therefore, if the strength of the internal rafter is reduced, there is a penalty to be paid in the design of the outer frame and this only emerges in the calculations when the equilibrium of the internal stanchions is considered.

There is no problem with the above design because of the order in which the decisions have been made and a well-balanced solution has been obtained with negligible bending moment in the internal stanchion. This is not necessarily always the case and the interested reader can try varying the design while still preserving the necessary equilibrium conditions for the reactant bending moment diagram.

With a balanced design, the internal stanchion is again little more than a simple prop. Nevertheless, although this member can be much lighter than the outer stanchions, it is recommended that a substantial member with a rigid connection to the rafter should be used in order to improve the stability of the structure as a whole and to take into account the possibility of unsymmetrical loading.

The influence of second-order effects is now checked in accordance with Sections 3.2.3 and 3.2.4 of this book. Bearing in mind that the second alternative design (with a non-symmetrical bending moment for the outer bay and the necessity to balance the bending moments in the internal stanchions) was introduced primarily for illustrative purposes, this check will be carried out for the first alternative design which is considered to be the practical solution. The bending moment diagram for this design is shown in Fig. 8.5 and the available load factor for rigid-plastic collapse is given approximately by

$$\lambda_p \approx \frac{761.25 + 245 - 228.7}{761.25} = 1.021$$

The member axial loads are

$$P_c = 125 \, \text{kN}$$
$$H = 459.65/6 = 76.6 \, \text{kN}$$
$$P_r = 76.6 \cos 5.71° + 125 \sin 5.71° = 88.7 \, \text{kN}$$

and the chosen members are

outer stanchion	457 × 191 × 74 UB	$I_c = 33\,430 \, \text{cm}^4$
rafter	406 × 140 × 46 UB	$I_r = 15\,670 \, \text{cm}^4$

so that

$$R = \frac{I_c L_r}{I_r h} = \frac{33\,430 \times 12\,562}{15\,670 \times 6000} = 4.467$$

$$\lambda_{cr} = \frac{3EI_r}{L_r \left[\left(1 + \dfrac{1.2}{R}\right) P_c h + 0.3 P_r L_r \right]}$$

$$= \frac{3 \times 205 \times 1.567 \times 10^8}{12\,562 \left[\left(1 + \dfrac{1.2}{4.467}\right) \times 125 \times 6000 + 0.3 \times 88.7 \times 12\,562 \right]}$$

$$= 5.97$$

and the required $\quad \lambda_p = \dfrac{0.9\lambda_{cr}}{\lambda_{cr} - 1} = \dfrac{0.9 \times 5.97}{4.97} = 1.081$

As this is greater than the available value of 1.021, a modest increase in the section sizes is necessary in order to satisfy the requirements of second-order sway stability effects. Alternatively, the effect of taking account of the stiffness of nominally pinned bases could be explored according to Chapter 3.

A check is also required on the influence of second-order effects on the stability of the internal rafters in accordance with Section 3.2.5 of this book. For this check, it is necessary to assume a size for the internal stanchions which, though rigidly connected to the rafter haunches, act primarily as struts carrying an axial load of 250 kN. A suitable member is:

$$\text{internal stanchion } 305 \times 165 \times 40\,\text{UB} \qquad I_c = 8551\,\text{cm}^4$$

so that

$$\lambda_{cr} = \left(\frac{D}{L_b}\right)\left[\frac{55\left(4 + \dfrac{L}{h}\right)}{\Omega - 1}\right]\left(1 + \frac{I_c}{I_r}\right)\left(\frac{275}{p_{yr}}\right)\tan 2\theta_r$$

$$= \left(\frac{402.3}{23\,000}\right)\left[\frac{55\left(4 + \dfrac{25\,000}{6000}\right)}{\cdot\,1.125 - 1}\right]\left(1 + \frac{8551}{15\,670}\right)\tan 11.42° = 19.6$$

where, in the above equation, the symbols are defined in Section 3.2.5 with

$$L_b = \text{effective span of the bay} = 25.0 - 2.0 = 23.0\,\text{m}$$

$$\Omega = \text{arching ratio} = \frac{10 \times (25 - 2 \times 2.0)^2}{16 \times 245} = 1.125$$

The above value of λ_{cr} is, of course, more than adequate.

8.2.2 Design for exceptional snow loading

This load case is shown in Fig. 8.3(b). The fact that the three spans are not equally loaded makes design for this case rather more complicated. Although it becomes apparent at an early stage that this case is not critical, the calculations will be completed in order to demonstrate the method. This again involves making the frame statically determinate by three cuts at the apices, as shown in Fig. 8.2.

Because of symmetry, it is only necessary to consider half of the structure. The equations for the free bending moments are:

C to B $M_x = 1.4x^2$

$M_C = 0$

$M'_B = 154.4\,\text{kNm}$

$\quad (x = 10.5\,\text{m at haunch})$

$M_B = 218.75\,\text{kNm}$

B to A $M_x = 218.75\,\text{kNm}$ $M_A = 218.75\,\text{kNm}$

C to D $M_x = 1.4x^2 + 0.1893x^3$ $M_C = 0$

$M'_D = 373.5\,\text{kNm}$

$\quad (x = 10.5\,\text{m at haunch})$

$M_D = 588.5\,\text{kNm}$

$\quad$ (in sub-frame CDF)

D to F $M_x = 588.5\,\text{kNm}$ $M_F = 588.5\,\text{kNm}$

$\quad$ (in sub-frame CDF)

E to D $M_x = 1.4x^2 + 0.1893x^3$ $M_E = 0$

$M'_D = 373.5\,\text{kNm}$

$\quad (x = 10.5\,\text{m at haunch})$

$M_D = 588.5\,\text{kNm}$

$\quad$ (in sub-frame EDF)

D to F $M_x = 588.5\,\text{kNm}$ $M_F = 588.5\,\text{kNm}$

$\quad$ (in sub-frame EDF)

These are shown in Fig. 8.8.

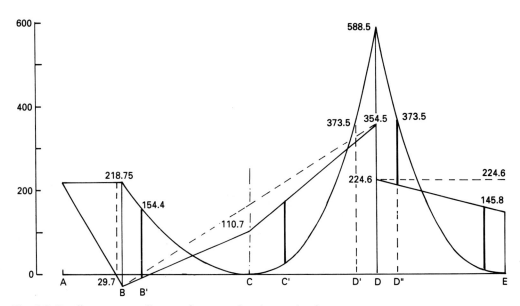

Fig. 8.8 Bending moment diagram for exceptional snow load case.

A reactant bending moment diagram is required for the side span and another for the centre span. It is important to ensure that the interaction between them results in zero bending moment at the pinned base of the internal stanchion.

A suitable diagram is shown in Fig. 8.8. This can be drawn by trial and error, taking care to satisfy the necessary equilibrium conditions. Alternatively, and perhaps more reliably, it can be drawn by solving the following equations which are based on the collapse mechanism shown in Fig. 8.9.

Noting that the sign convention is tension on the outside positive *in the individual sub-frames* and that symmetry requires that $V_2 = 0$:

At A	0	$= -M_1$	$-7.25H_1$	$+12.5V_1$	$+218.75$
At B'	M_{pr}	$= -M_1$	$-1.05H_1$	$+10.5V_1$	$+154.4$
At C'	$-M_{pr}$	$= -M_1$	$-0.40H_1$	$-4.0V_1$	$+34.5$
At F in CDF	M_F	$= -M_1$	$-7.25H_1$	$-12.5V_1$	$+588.5$
At F in EDF	M_F	$= -M_2$	$-7.25H_2$		$+588.5$
At D''	M_{pr}	$= -M_2$	$-1.05H_2$		$+373.5$
At E'	$-M_{pr}$	$= -M_2$	$-0.2H_2$		$+7.1$

These equations are not difficult to solve although a certain amount of trial and error may still be necessary in order to achieve acceptable plastic hinge positions in the rafters.

The solution which was used to draw the reactant moment lines in Fig. 8.8 is based on slightly more accurate plastic hinge positions than those shown in Fig. 8.9 and is:

$$M_1 = 110.7\,\text{kNm} \qquad M_2 = 145.8\,\text{kNm}$$
$$H_1 = 41.4\,\text{kNm} \qquad H_2 = 63.0\,\text{kNm}$$
$$V_1 = 15.4\,\text{kNm}$$
$$M_{pr} = 161.6\,\text{kNm}$$

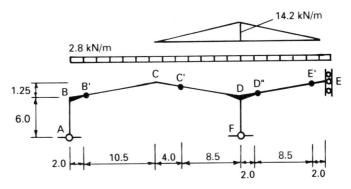

Fig. 8.9 Assumed collapse mechanism for exceptional snow load case.

This value of M_{pr} is, of course, considerably less than the value of 245 kNm used to draw Fig. 8.5 for the symmetrical snow load case, a fact that could have been anticipated once the magnitudes of the free bending moments were compared. Indeed, in practice, it may be more convenient to carry out an elastic analysis in such cases in order to demonstrate that the frame is elastic under the ultimate limit state loads and that plastic analysis is unnecessary.

8.3   Example 8.2. Three-span portal frame supported on valley beams

This example is virtually the same as Example 8.1, as shown in Fig. 8.3, except that alternate frames are supported on valley beams along lines D and G.

8.3.1   Symmetrical snow loading

The design for symmetrical snow loading shown in Fig. 8.5 for the conventional portal frame is also applicable to the comparable frame when the internal valleys are supported on valley beams. This is because the rafter thrusts and bending moments balance at the top of the internal stanchions and there are no out of balance horizontal forces or bending moments (applied as torsional moments) applied to the valley beam.

However, in the valley beam frames, second-order effects can be more severe and they should be checked in accordance with Section 3.2.6 of this book:

From Example 8.1, the member axial loads are:

$P_{\mathrm{c}} = 125\,\mathrm{kN}$
$H = 459.65/6 = 76.6\,\mathrm{kN}$
$P_{\mathrm{r}} = 76.6\cos 5.71° + 125\sin 5.71° = 88.7\,\mathrm{kN}$

and the member sizes are:

| outer stanchion | $457 \times 191 \times 74\,\mathrm{UB}$ | $I_{\mathrm{c}} = 33\,430\,\mathrm{cm}^4$ |
| rafter | $406 \times 140 \times 46\,\mathrm{UB}$ | $I_{\mathrm{r}} = 15\,670\,\mathrm{cm}^4$ |

so that:

$$R = \frac{I_{\mathrm{c}}L_{\mathrm{r}}}{I_{\mathrm{r}}h} = \frac{33\,430 \times 12\,562}{15\,670 \times 6000} = 4.467$$

$$\lambda_{cr} = \frac{3EI_r}{2L_r\left[\left(1 + \frac{1.2}{R}\right)P_ch + 0.6P_rL_r\right]}$$

$$= \frac{3 \times 205 \times 1.567 \times 10^8}{2 \times 12\,562\left[\left(1 + \frac{1.2}{4.467}\right) \times 125 \times 6000 + 0.6 \times 88.7 \times 12\,562\right]}$$

$$= 2.37$$

This value of the elastic critical load λ_{cr} is very much less than the limiting value of 4.6 (see Section 3.2.7) and therefore the frame is unacceptable unless proven satisfactory by means of an elastic-plastic second-order analysis of the complete frame.

A review of the formula indicates that for λ_{cr} to be increased to 4.6 then the rafter I_{xx} would have to be approximately doubled and this is likely to be commercially unacceptable. Other means of *stiffening* the frame should therefore be considered such as bracing the valley beam frames back to the valley stanchion frames. This stiffening could be triangulated bracing in the roof plane which incorporates the valley beam itself as part of the bracing system or, alternatively, BS 5950: Part 9 could be used to justify that the roof cladding acting as a diaphragm can provide the required additional stiffness.

An examination of the calculation for internal rafter stability given in Section 8.2.1 for Example 8.1 shows that again this is not critical. If the value of I_c for the internal stanchion is taken to be zero, $\lambda_{cr} = 19.6/1.546 = 12.7$ which is greater than the limiting value of 10.

8.3.2 Exceptional snow loading

Under exceptional snow loading, the final design for Example 8.1 shown in Fig. 8.8 is not appropriate because it puts torsional moments into the valley beam. It is, therefore, necessary to recalculate the reactant bending moment diagram in order to obtain bending moments and horizontal thrusts which balance in the valleys.

The resulting bending moment diagram is shown in Fig. 8.10. The corresponding values of the redundant forces are:

$$
\begin{array}{ll}
M_1 = 38.1\,\text{kNm} & M_2 = 168.6\,\text{kNm} \\
H_1 = 42.9\,\text{kN} & H_2 = 42.9\,\text{kN} \\
V_1 = 10.4\,\text{kN} & \\
M_{pr} = 180.8\,\text{kNm} &
\end{array}
$$

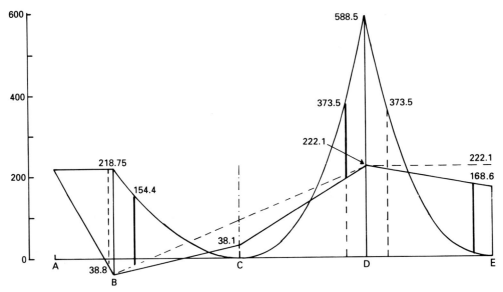

Fig. 8.10 Bending moment diagram for frame supported on valley beams.

8.4 Example 8.3. Two-span portal frame with unequal spans

This example, together with its required ultimate limit state loading, is shown in Fig. 8.11.

Once again, the frame is made statically determinate by introducing cuts at C and E giving rise to six redundant forces H_1, V_1, M_1, H_2, V_2, M_2. The free bending moments are then very much as in the previous examples, and the equation for each of the rafters is

$$M_x = 5x^2$$

which gives rise to the free bending moment diagram shown in Fig. 8.13.

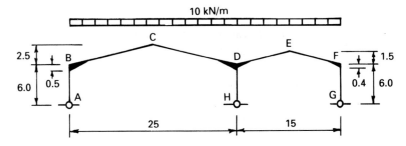

Fig. 8.11 Two-span portal frame.

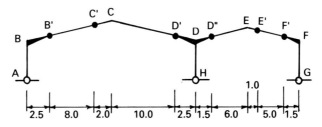

Fig. 8.12 Assumed collapse mechanism for the two-bay frame.

In order to draw the corresponding reactant moment diagram, it is necessary to make some assumptions regarding the collapse mechanism and the one shown in Fig. 8.12 will be used. This takes note of the fact that it is again impossible to avoid plastic hinges at the ends of the valley haunches and, if hinges are assumed at the ends of all of the haunches, $V_1 = V_2 = 0$ and the reactant lines for the rafters are horizontal. Equilibrium in the internal stanchion is also automatically ensured.

The plastic hinge positions shown in Fig. 8.12 for the apex hinges are, of course, estimates but small changes in these hinge positions will not have much effect on the solution and they can be adjusted later if necessary. The easiest way to draw the reactant moment diagram corresponding to this mechanism is probably to solve the equilibrium equations, thus.

For the left-hand bay

$$
\begin{array}{llll}
\text{At A} & 0 & = -M_1 & -8.5H_1 & +781.25 \\
\text{At B' and D'} & M_{\text{pr}} & = -M_1 & -2.0H_1 & +500 \\
\text{At C'} & -M_{\text{pr}} & = -M_1 & -0.4H_1 & +20 \\
\text{which gives} & H_1 & = 71.4\,\text{kN} \\
& M_1 & = 174.3\,\text{kNm} \\
& M_{\text{pr}} & = 182.9\,\text{kNm}
\end{array}
$$

and the reactant line for the left-hand bay in Fig. 8.13 is drawn on this basis. A suitable choice of rafter is a 406 × 140 × 39 UB with a full plastic moment of 198 kNm. This gives a load factor of $198/182.9 = 1.083$ and it will be checked later whether this is sufficient to overcome second-order effects.

The bending moment below the haunch determines the size of the stanchion and this is $71.40 \times 5.5 = 392.7\,\text{kNm}$. A suitable section is therefore a 457 × 191 × 67 UB with a full plastic moment of 405 kNm.

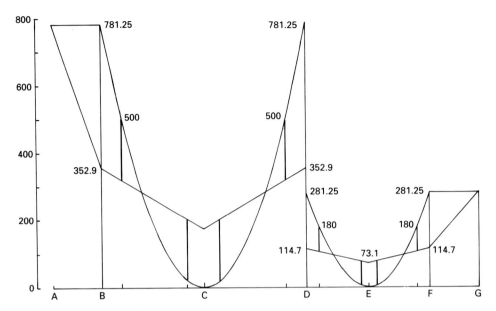

Fig. 8.13 Bending moment diagram for two-bay frame.

For the right-hand bay

$$
\begin{array}{llll}
\text{At G} & 0 & = -M_2 & -7.5H_2 & +281.25 \\
\text{At D' and F'} & M_{pr} & = -M_2 & -1.2H_2 & +180 \\
\text{At E'} & -M_{pr} = -M_2 & -0.2H_2 & +5 \\
\text{which gives} & H_2 & = 27.8\,\text{kN} \\
& M_2 & = 73.1\,\text{kNm} \\
& M_{pr} & = 73.6\,\text{kNm}
\end{array}
$$

and the reactant line for the right-hand bay in Fig. 8.13 is drawn on this basis. A suitable choice of rafter is a $305 \times 102 \times 25$ UB with a full plastic moment of $92.4\,\text{kNm}$. This gives a load factor of $92.4/73.6 = 1.255$ to overcome second-order effects.

The bending moment below the haunch again determines the size of the stanchion and this is $27.8 \times 5.6 = 155.4\,\text{kNm}$. A suitable section is therefore a $356 \times 127 \times 39$ UB with a full plastic moment of $180\,\text{kNm}$.

The influence of second-order effects is now checked in accordance with Section 3.2.4 of this book. Because the frame is irregular, it is necessary to check both outer bays. For the left-hand bay:

The member axial loads are

$$
\begin{aligned}
P_c &= 125\,\text{kN} \\
H &= 71.4\,\text{kN} \\
P_r &= 71.4\cos 11.3° + 125\sin 11.3° = 94.5\,\text{kN}
\end{aligned}
$$

and the chosen members are

outer stanchion $457 \times 191 \times 67$ UB $I_c = 29\,410\,\text{cm}^4$
rafter $406 \times 140 \times 39$ UB $I_r = 12\,410\,\text{cm}^4$

so that

$$R = \frac{I_c L_r}{I_r h} = \frac{29\,410 \times 12\,748}{12\,410 \times 6000} = 5.035$$

$$\lambda_{cr} = \frac{3EI_r}{L_r \left[\left(1 + \frac{1.2}{R} \right) P_c h + 0.3 P_r L_r \right]}$$

$$= \frac{3 \times 205 \times 1.241 \times 10^8}{12\,748 \left[\left(1 + \frac{1.2}{5.035} \right) \times 125 \times 6000 + 0.3 \times 94.5 \times 12\,748 \right]}$$

$$= 4.64$$

and the required $\lambda_p = \dfrac{0.9\lambda_{cr}}{\lambda_{cr} - 1} = \dfrac{0.9 \times 4.64}{3.64} = 1.147$

The above value of λ_{cr} is only just above the limiting value of 4.6 below which a more precise treatment of second-order effects becomes mandatory. Bearing in mind that the approximate treatment is conservative, in this case it is better to use a more exact treatment if at all possible. This could advantageously take account of nominal partial base fixity. According to the approximate method, as the required value of λ_{cr} is greater than the available value of 1.083, an increase in the section sizes of the left-hand bay is necessary in order to satisfy the requirements of second-order sway stability effects.

Similarly, in the right-hand bay:

The member axial loads are

$P_c = 75\,\text{kN}$
$H = 27.8\,\text{kN}$
$P_r = 27.8 \cos 11.3° + 75 \sin 11.3° = 41.9\,\text{kN}$

and the chosen members are

outer stanchion $356 \times 127 \times 39$ UB $I_c = 10\,100\,\text{cm}^4$
rafter $305 \times 102 \times 25$ UB $I_r = 4364\,\text{cm}^4$

so that

$$R = \frac{I_c L_r}{I_r h} = \frac{10\,100 \times 7649}{4364 \times 6000} = 2.95$$

$$\lambda_{cr} = \frac{3EI_r}{L_r\left[\left(1 + \frac{1.2}{R}\right)P_c h + 0.3P_r L_r\right]}$$

$$= \frac{3 \times 205 \times 0.4364 \times 10^8}{7649\left[\left(1 + \frac{1.2}{2.95}\right) \times 75 \times 6000 + 0.3 \times 41.9 \times 7649\right]} = 4.81$$

and the required $\lambda_p = \dfrac{0.9\lambda_{cr}}{\lambda_{cr} - 1} = \dfrac{0.9 \times 4.81}{3.81} = 1.136$

Again, the above value of λ_{cr} is only just above the limiting value of 4.6 below which a more precise treatment of second-order effects becomes mandatory. However, here the available reserve of safety in the preliminary design gave $\lambda_p = 1.255$ so that no increase in section sizes is required.

The bending moment below the haunch in the internal stanchion is $392.7 - 152.9 = 239.8\,\text{kNm}$. Possibly, for design, this should be factored up by $\lambda_p = 1.147$ to give $275.1\,\text{kNm}$ though this is debatable and need not be insisted upon if the design were tight. A suitable section to carry this moment together with an axial load of $125 + 75 = 200\,\text{kN}$ is a $457 \times 152 \times 52$ UB in design Grade 43 which has a gross plastic moment of $301\,\text{kNm}$ and a net plastic moment of $294\,\text{kNm}$. However, this section size is subject to there being sufficient restraints to ensure member stability. Clients do not always accept intermediate restraints to internal stanchions because they reduce access between spans and, if this were the case, a larger member would be necessary dependent on the restraint available.

8.5   Example 8.4. Buttressed multi-span portal frame

This example, which is subject to a vertical uniformly distributed load of $10\,\text{kN/m}$ at the ultimate limit state, is shown in Fig. 8.14. From the point of view of the plastic design, the number of spans is not important as the design forces in the internal spans are all identical.

The main principle in the buttress frame is to design the inner spans with relatively light rafters taking advantage of a high arching ratio and to design each side span as a buttress that has to hold up the internal

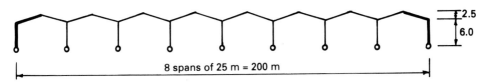

2.5

6.0

8 spans of 25 m = 200 m

Fig. 8.14 Eight-bay portal structure with outer buttress bays.

spans. In addition to possible economic benefits, the advantages of this are reduced eaves spread and no plastic hinges in the eaves zone of the side stanchions early in the hinge history. Another benefit is that temperature changes in the frame tend to cause the apices to move vertically rather than for all of the temperature movement to appear as eaves spread.

There are various alternative ways of designing buttress frames. In the example which follows, the buttress frame is a pinned based structure with a larger side stanchion and with a larger rafter in the outer half span of the outer frame. An alternative is to use fixed base outer and inner stanchions with larger rafters in both halves of the span. A fixed base for the outer stanchions is particularly advantageous. The choice will depend on the frame size, geometry and loading and the performance criteria for deflections.

The buttress effect results in a significant increase in the axial compressive force in the internal rafters so that it is particularly important to check these for 'snap-through' as covered in BS 5950: Part 1, clause 5.5.3.3 or Section 3.2.5 of this book.

Conversely, the buttress frames will be stiff and will be relatively less affected by second-order sway stability effects. However, unless more accurate calculations can be made, they should be checked by adapting the procedures given in Sections 3.2.3 and 3.2.4 of this book.

This example only gives guidance on the procedures that may be used in order to make an initial choice of sections. The further working out of the design is then left to the interested reader. There is infinite scope for trial and error adjustment of the design in order to obtain a good fit with the available section sizes and the various constraints on the design.

The internal spans are chosen first. We may observe here that if a balanced design were to be used, without taking advantage of buttress action, the initial calculation would be identical to that used for the left-hand span of the two-span frame considered in the previous section. Ignoring the possible influence of second-order effects, this would result in a requirement for the full plastic moment of the rafters of $M_{pr} = 182.9\,\text{kNm}$ with a horizontal thrust of $H = 71.4\,\text{kN}$.

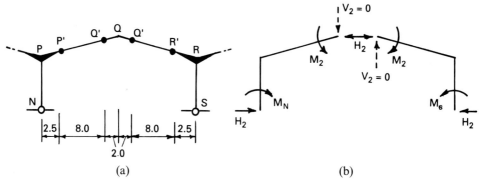

Fig. 8.15 Plastic hinges for the internal span of a buttress frame: (a) plastic hinges; (b) redundant force system.

We will assume that buttress action allows us to reduce M_{pr2} in the internal rafters to 150 kNm and that a suitable section, with a full plastic moment a little greater than this to allow for second-order effects, is available. If we make each internal span statically determinate by means of an apex cut, as before, the bending moment diagrams for all of the internal spans will be identical with $V_2 = 0$. If we further assume the plastic hinge positions shown in Fig. 8.15 (which also require plastic hinges in the outer spans before they constitute a valid mechanism), then:

$$\begin{aligned}
\text{At } Q' && -150 &= -M_2 && -0.4H_2 && +20 \\
\text{At } P' \text{ and } R' && +150 &= -M_2 && -2.0H_2 && +500 \\
\text{giving} && H_2 &= 112.5\,\text{kN} \\
&& M_2 &= 125.0\,\text{kNm}
\end{aligned}$$

so that even a relatively modest reduction in M_{pr2} from 182.9 to 150 kNm requires a significant increase in H_2 from 71.4 to 112.5 kN with a consequential increase in the strength of the buttressing members together with an increased tendency towards buckling of the internal rafters. It follows that this procedure has to be used with care.

The bending moment diagram arising from this calculation is shown in the right-hand part of Fig. 8.16. Because we have not made any attempt to control the bending moment at the pinned bases of the penultimate stanchions, the implied bending moments have to be calculated and balanced by the corresponding bending moments in the outer buttress frames.

Thus, at N, $M_{\mathrm{N}} = -M_2 - 8.5H_2 + 781.25 = -300.0\,\text{kNm}$

For the outer bay, with the benefit of a little experience or trial and error, we may assume the plastic hinge positions shown in Fig. 8.17

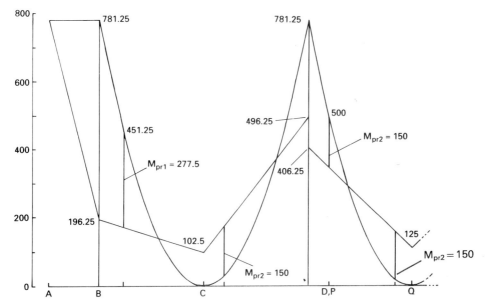

Fig. 8.16 Bending moment diagram for the outer spans of the buttress frame.

which, together with the hinges shown in Fig. 8.15, constitute a valid collapse mechanism for the frame. Noting that the bending moment at E must balance $M_N = -300\,\text{kNm}$ calculated above, this leads to the following equations which include the required full plastic moment M_{pr1} of the outer rafter as an unknown.

$$
\begin{array}{lrcrrrr}
\text{At A} & 0 & = -M_1 & -8.5H_1 & +12.5V_1 & +781.25 \\
\text{At B}' & M_{pr1} = & -M_1 & -1.9H_1 & +9.5V_1 & +451.25 \\
\text{At C}' & -150 & = -M_1 & -0.5H_1 & -2.5V_1 & +31.25 \\
\text{At E} & -300 & = -M_1 & -8.5H_1 & -12.5V_1 & +781.25
\end{array}
$$

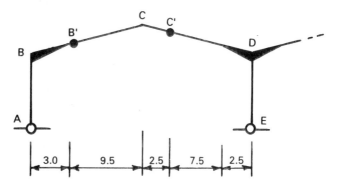

Fig. 8.17 Plastic hinge positions in the outer span.

The first and last of these equations lead directly to $V_1 = 12.0\,\text{kN}$ and the remaining equations are then easily solved to give:

$$
\begin{aligned}
M_1 &= 102.5\,\text{kNm} \\
H_1 &= 97.5\,\text{kN} \\
V_1 &= 12.0\,\text{kN} \\
M_{\text{pr1}} &= 277.5\,\text{kNm}
\end{aligned}
$$

The remainder of the bending moment diagram shown in Fig. 8.16 can now be drawn.

The bending moment at B is $781.25 - 196.25 = 585\,\text{kNm}$ so that the design bending moment for the outer stanchion below the haunch is $585 \times 5.5/6.0 = 536.3\,\text{kNm}$.

The design moment below the haunch in the penultimate stanchion is $(496.25 - 406.25) \times 5.5/6.0 = 82.5\,\text{kNm}$. This gives a complete pattern of bending moments at collapse which can be used as the basis for a choice of members. The procedure used allows for easy adjustment in order to accommodate the available member sizes and the constraints of second-order effects, deflections, etc.

An initial choice of sections can be based on the bending moment diagram shown in Fig. 8.16. The internal rafters can be $406 \times 140 \times 39\,\text{UB}$ with a full plastic moment of $198\,\text{kNm}$ compared with the minimum of $150\,\text{kNm}$ required. The reduction in bending capacity due to the axial load is of the order of 1% and can be ignored. A check on second-order effects can be made assuming a $406 \times 178 \times 54\,\text{UB}$ internal stanchion:

$$
L_b = \text{effective span of the bay} = 25.0 - 2.5 = 22.5\,\text{m}
$$

$$
\Omega = \text{arching ratio} = \frac{10 \times (25 - 2 \times 2.5)^2}{16 \times 195.6} = 1.278
$$

so that

$$
\lambda_{\text{cr}} = \left(\frac{D}{L_b}\right) \frac{\left[55\left(4 + \dfrac{L}{h}\right)\right]}{\Omega - 1} \left(1 + \frac{I_c}{I_r}\right) \left(\frac{275}{p_{\text{yr}}}\right) \tan 2\theta_r
$$

$$
= \left(\frac{397.3}{22\,500}\right) \frac{\left[55\left(4 + \dfrac{25\,000}{6000}\right)\right]}{1.278 - 1} \left(1 + \frac{18\,670}{12\,410}\right) \tan 22.6° = 29.7
$$

and the internal span sections are satisfactory.

The outer span can comprise

Outer stanchion:
 Requires $M_{ps} = 536.3\,\text{kNm}$ $533 \times 210 \times 92\,\text{UB}$ gives $651\,\text{kNm}$
Outer rafter:
 Requires $M_{pr1} = 277.5\,\text{kNm}$ $457 \times 152 \times 60\,\text{UB}$ gives $353\,\text{kNm}$
Inner rafter:
 Requires $M_{pr2} = 150.0\,\text{kNm}$ $406 \times 140 \times 39\,\text{UB}$ gives $198\,\text{kNm}$

The check for second-order effects in the outer spans then gives:

$$H = 97.5\,\text{kN}$$
$$P_c = 125 + 12 = 137.0\,\text{kN}$$
$$P_r = H\cos\theta + P_c\sin\theta = 122.5\,\text{kN}$$

$$R = \frac{I_c L_r}{I_r h} = \frac{55\,330 \times 12\,748}{25\,450 \times 6000} = 4.62$$

$$\lambda_{cr} = \frac{3EI_r}{L_r\left[\left(1 + \dfrac{1.2}{R}\right)P_c h + 0.3 P_r L_r\right]}$$

$$= \frac{3 \times 205 \times 2.545 \times 10^8}{12\,748\left[\left(1 + \dfrac{1.2}{4.62}\right) \times 137 \times 6000 + 0.3 \times 122.5 \times 12\,748\right]}$$

$$= 8.16$$

The required λ_p is then $\dfrac{0.9 \times 8.16}{8.16 - 1} = 1.026$

It follows that the minimum requirements for the full plastic moments then become

Outer stanchion: $M_{ps} = 1.026 \times 536.3 = 550.1\,\text{kNm}$
Outer rafter: $M_{pr1} = 1.026 \times 277.5 = 284.6\,\text{kNm}$
Inner rafter: $M_{pr2} = 1.026 \times 150.0 = 153.9\,\text{kNm}$

and the initial sections are adequate for further detailed checks.
 It may be noted here that the above calculation of λ_{cr} is based on the stiffest outer members of a frame which includes members of disparate stiffness. It could be argued that this is potentially unconservative and, when other assumptions are tried, it is found that the value of λ_{cr} is quite sensitive to the assumptions made. In such cases, the use of a computer program to calculate an accurate value of λ_{cr} could be beneficial.

References

8.1   Macleod I A. Guidelines for checking computer analysis of building structures. CIRIA Technical Note 133, 1988.

8.2   Code of practice for stressed skin design, Part 9 of BS 5950: Structural use of steelwork in building, 1994.

8.3   Davies J M. Plastic collapse of framed structures clad with corrugated steel sheeting. *Proc. ICE*, Part II, Vol. 55, March 1973.

Chapter 9
Design of Agricultural Buildings

9.1 Introduction

In principle, agricultural buildings are no different from any other low-cost low-rise buildings and can be designed using identical procedures. As agricultural buildings are frequently steel portal frames or similar structures, the relevant parts of the first eight chapters of this book are directly applicable. However, there are some additional factors to be considered and the purpose of this chapter is to outline some of these and the influence that they may have on the design.

Agricultural buildings tend to be smaller in size than industrial buildings and designed to less demanding specifications. The main considerations to be addressed are therefore:

- smaller design loads
- smaller spans and steeper roof pitch
- simpler details
- loads from stored crops.

As climatic loads, such as wind and snow, are unaware whether they are being applied to an agricultural or a conventional building, the use of smaller design loads implies reduced factors of safety. It is, therefore, particularly important to ensure that there is not a further erosion of the safety level because of inadequate detailing or a cavalier attitude to stability checks. Unfortunately, this is not always the case.

9.2 Design code for agricultural buildings

The reason why smaller design loads may be applied to agricultural buildings is the existence of a British Standard[9.1] which is specific to this class of structures. BS 5502, however, merely adjusts the design load and, in particular, it does not change any of the design procedures embraced in BS 5950.

BS 5502 has numerous Parts and the structure of the complete standard is described in Part 0: Introduction. However, only Part 20: Code of Practice for general design considerations and Part 22: Code of Practice for Design, Construction and Loading are directly relevant to the subject of this book.

9.2.1   Building classification

BS 5502: Part 22: 1993 introduces the concept of design classification according to the density of human occupancy, the location of the building and the return period (design life) of the loadings. It should be noted that, in this context, the design life is concerned solely with the statistical likelihood of high levels of load being achieved and is not directly related to the durability of the materials of construction. Thus, in buildings which have a low level of human occupancy, the consequences of structural failure are less serious in terms of danger to life and such buildings can be designed to lower levels of load. The design classifications are given in Table 9.1 where the risk of collapse is greater in buildings or structures which have the higher classification numbers.

Table 9.1 Design classification of buildings and structures for agriculture.

Class	Maximum normal human occupancy within a building or structure or its zone of effect where applicable (person h/year)	Minimum allowable distance to a classified highway or human habitation (m)	Minimum design life (years)
1	Unrestricted	Unrestricted	50
2	Not exceeding 6 h/day at a maximum density of 2 persons/50 m² (4380/50 m²)	Either 10 or limit of zone of effect if greater than 10	20
3	Not exceeding 2 h/day at a maximum density of 1 person/50 m² (730/50 m²)	Either 20 or limit of zone of effect if greater than 20	10
4	Not exceeding 1 h/day at a maximum density of 1 person/50 m² (365/50 m²)	Either 30 or limit of zone of effect if greater than 30	2

Table 9.2 Snow load conversion factor ω.

Building class	ω
1	1.00
2	0.78
3	0.61
4	0.22

Having classified the building or structure, the characteristic imposed roof loads may then be modified from those given in BS 6399: Part 3: 1988 according to either (a) or (b) as follows:

(a) For class 1 buildings and structures and for classes 2, 3 and 4 where there is access to the roof, the minimum imposed load on a roof should be in accordance with BS 6399. For buildings of classes 2, 3 and 4 where there is no access to the roof except for cleaning and repair, the minimum uniformly distributed load should be taken as 0.3 kN/m^2 measured on plan for roof slopes of less than 60°, or as zero for roof slopes greater than or equal to 60°.

Snow loads given in BS 6399 should be assumed to be applicable to class 1 buildings. For other classifications, the derived snow load on the roof (S_d) should be modified by multiplying by the factor ω given in Table 9.2.

(b) Provided that the altitude is less than 100 m and that access to the roof is restricted to that necessary for cleaning and repair, the uniformly distributed load given in Table 9.3 is applied symmetrically to the full plan area of the roof.

The wind loads given in CP3: Chapter V: Part 2: 1972 are not modified.

Table 9.3 Generalised imposed roof load.

Building class	Load (kN/m^2)
1	0.64
2	0.50
3	0.40
4	as in (a)

Note: Care should be taken in using this table where localised conditions of climatological effects or drifting (e.g. where there are abrupt changes of building height) may give rise to higher loadings.

9.3   Design with simpler details

This section briefly discusses the consequences of some typical current practices in the design of agricultural buildings. These follow from the illogical argument that lower levels of load can be associated with simpler details. A typical example of current practice is the omission of fly braces, as shown in Fig. 3.27, from the rafters of small portal frames. This assumes that the rafter members are sufficiently stocky for the purlins and purlin cleats to provide adequate restraint to the bottom flange. This is certainly possible in the case of the RSJ rafters which were used in the first of the case studies in Chapter 10. It is certainly not possible when relatively slender Universal Beams are used and the designer needs to be cautious here and not to omit fly braces without proper consideration of member stability. It is particularly important to note that the purlin to cleat to rafter connection requires at least two bolts if it is to provide torsional restraint to the rafter.

Another frequent simplification with agricultural buildings is the omission of shear and bearing stiffeners in connections, particularly the haunch connections in pitched roof portal frames. Here again, this *may* be possible for frames of small span but it is not generally so. It is shown in the next chapter that this can lead to a significant reduction in the carrying capacity of a typical frame of small span.

Small columns have relatively thin flanges and this can make it necessary to provide a cap plate at the top of the column in order to stiffen and strengthen the flange adjacent to the bolts in tension. If this is not done, one consequence of the deformation of the flange will be increased deflection at the apex and a likely decrease in the carrying capacity as a result of enhanced second-order effects. Another consequence is a reduction of the lever arm of the tension bolt group with a consequential increase in the shear and bearing stresses in the top of the column for a given bending moment. These are further reasons why the use of lightweight construction must not be allowed to reduce the attention to detail in the design.

The bases of the smaller columns that are frequently used in agricultural buildings can be conveniently fixed by placing them in a socket in the concrete base block and relying on 'garden spade' action to provide the fixity. Guidance on the necessary foundation design can be found in the *Steel Designers Manual*[9.2]. Thus, the fixed base detail is often more appropriate to farm buildings than to the larger industrial buildings. However, this method requires considerably more concrete than in a comparable pinned base design and is only applicable when the ground can provide adequate bearing pressure.

The roof cladding of agricultural buildings is frequently fibre-reinforced cement sheets fastened to the supporting purlins with hook bolts through the crests of the profile. This is perfectly good practice but it does mean that such buildings do not have the benefit of any significant 'stressed skin' effect which is available to provide additional stability in the case of conventional metal cladding fastened through the troughs. This is yet another reason for not taking too many liberties with the stability of agricultural buildings.

9.4   Loads from stored crops

Agricultural buildings are frequently designed to store crops and other bulk materials as illustrated by Fig. 9.1. The various parts of BS 5502, and notably Table 7 of BS 5502: Part 22: 1993, give a good deal of information regarding the loads imposed on the structure by the various materials that may be stored.

This has two design consequences which should not be overlooked. The active pressure from the stored material leaning against the side of the building applies a significant point load to the frame at the height of the horizontal rail. This load tends to crush the web of the stanchion and this web should be checked for buckling and bearing. Second, and more important, the point load modifies the shape of the bending moment diagram, as shown in Fig. 9.2. The large region of approximately uniform bending moment above the load point may cause a significant reduction in the stability of the stanchion. In an extreme case, this entire region may be fully plastic at or near the collapse load and its stability will therefore require to be carefully checked in order to ensure that the design load can be achieved.

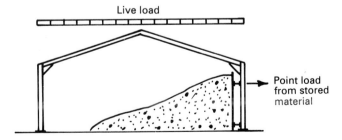

Fig. 9.1 Storage of bulk materials.

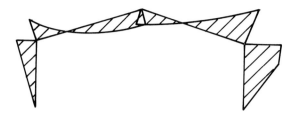

Fig. 9.2 Bending moment diagram with stored materials.

Furthermore, the horizontal point load from stored crops will cause a horizontal deflection at the eaves which may well be the design criterion rather than strength. Conversely, the required increase in member sizes, coupled with a reduced axial compression in the rafter, reduces the sensitivity of structures loaded in this way to second-order effects.

The consequences of the load from stored material on the performance of a typical agricultural building frame are explored in some detail in the next chapter.

References

9.1 British Standards Institution. BS 5502 Buildings and Structures for Agriculture.

9.2 *Steel Designers Manual*, 5th ed. Steel Construction Institute and Blackwell Scientific Publications, 1994.

Chapter 10

Performance of Agricultural Buildings – Two Case Studies

10.1 Introduction

A major research project was recently undertaken by the Universities of Manchester and Salford in which two 12 m span pitched roof portal frames were designed to the Agricultural Code, built full size in the laboratory, and tested to destruction. It is significant that both frames were designed by practitioners outside the academic research project and that both incorporated design faults that resulted in a reduction in the carrying capacity. This chapter describes these tests and the lessons that can be learned from them.

10.2 General arrangement for the tests

The test procedures have been described in References 10.1 and 10.2 and the description of the first test included in this chapter is taken largely from Reference 10.2.

One of the fundamental decisions taken at the planning stage was that the tests should simulate, as realistically as possible, the actual load and restraint conditions existing in a portal frame within an actual structure. This meant that the 'snow' loading would have to be applied through the sheeting and thence onto the purlins. A three-dimensional test assembly was therefore developed which allowed a realistic degree of structural interaction between components to be reproduced. In particular, the restraint offered by secondary members to the main frame was modelled as accurately as possible. This was important, as one of the primary areas of interest was the lateral stability of members, particularly the haunched region.

It was necessary to design the complete test assembly to fit into the available area of strong floor in the Structures Laboratory at the University of Salford (approximately 24 m × 10 m). Figure 10.1 shows

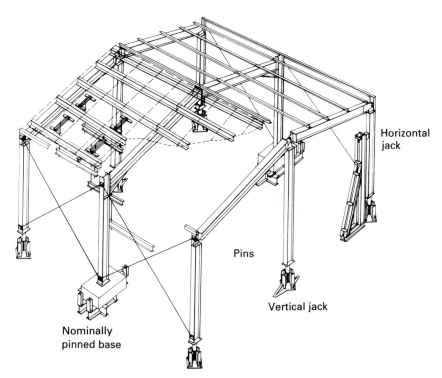

Fig. 10.1 General arrangement for portal frame tests.

the test rig which consisted of three 12 m span frames with the central test frame connected at rafter level by cold-formed purlins to the two 'gable' frames. Lateral movement of the gable frames was restrained by bracing them back to the stanchions in the laboratory walls.

The load was applied to the sheeting by hydraulic jacks through a system of beams, hangers and timber spreaders. Figure 10.2 is an interior view of the test set-up showing the arrangement of a hydraulic jack together with its associated beams and hangers positioned below the roof level, while Fig. 10.3 is an exterior view indicating the placement of the timber load-spreading members on the roof sheeting. Each spreader applied 6 load points locally to the sheeting so that a total of 96 load points were controlled by each jack to give a total of 384 load points for the 12 m span frames. The loading system was designed to have a maximum test capacity of $2\,kN/m^2$. This, together with the self-weight of the frame, purlins, sheeting and the loading system itself (approximately $1\,kN/m^2$), was sufficient to ensure failure of the frames to be tested.

Fig. 10.2 Interior view with loading system.

Fig. 10.3 Exterior view with roof loads.

If the two gable frames had been made rigid, loading through the roof sheeting and purlins could have given rise to parasitic stressed skin effects which would have significantly influenced the behaviour of the test frames. Considerable thought was given to methods of avoiding this, and it was finally decided that the gable frames should be fully articulated by arranging that the intersecting members of the gable frames were pin-jointed at the eaves and apices and that all the gable posts were located on knife-edge supports. This allowed the nodal deflections of these frames to be controlled precisely by means of two independent jacks. The principles involved can be visualised with the aid of Fig. 10.1. A hydraulic jack was positioned beneath the central gable post, and this controlled the vertical movement of the apex joint. The horizontal movement of the gable eaves was also controlled by a push-pull jack, as shown in Fig. 10.4. At each load increment, the nodal deflections of the test frame were measured and the articulated side frames were then adjusted until their corresponding nodal deflections coincided with those of the test frame.

A secondary effect that could not be satisfactorily eliminated was the partial continuity of the purlins. This resulted in the central frame receiving more load than the two gable frames. The precise distribution of load on the frame could not be readily assessed by simple analysis, as it later became apparent that the distribution was varying during the tests. It was necessary, therefore, to measure accurately the gravity load

Fig. 10.4. Eaves detail of gable frame.

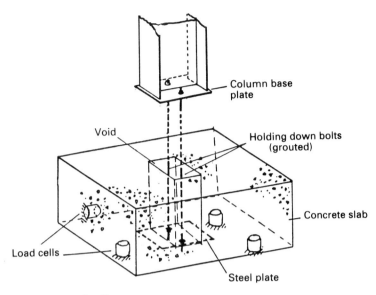

Fig. 10.5. Base detail.

being applied to the frames at each stage during the test and this was achieved by means of load cells positioned below the concrete base slabs of the test frame.

It was decided that not only the loading system and restraints but also the base detail should be realistic. The frames were therefore detailed to have nominally 'pinned' bases sitting on concrete foundation blocks. Figure 10.5 shown the general arrangement of these bases and the positions of the four load cells used to monitor the horizontal load, the vertical load, and the bending moments acting on the concrete base. The bases were designed to be reused so they were cast with a central rectangular opening into which holding-down bolts could be grouted in a separate operation.

In addition to the load cells mentioned above, the frames were fully instrumented by means of some 180 strain gauges and 20 deflection transducers. Data collection and processing was fully automated by means of data logging equipment and interlinked computers.

10.3   Details of Frame 1

The fabricated details of the first frame are shown in Fig. 10.6. The frame was designed as an agricultural building to BS 5502[10.3] with an imposed load of 0.462 kN/m^2 and a load factor of 1.7. As with each of the frames that were tested in this project, the design and detailing were undertaken by an experienced designer/fabricator within the steel construction industry without any interference from the academics directing the project, other than specifying the physical limitations of the laboratory space.

It is interesting to note at the outset that each of the two tested frames had at least one poor design feature which caused some form of premature failure. In each case, adherence to the guidance given in BS 5950: Part 1 would have eliminated the problem. In the case of Frame 1, the poor design features were twofold.

(1)   the web of the column adjacent to the haunch was overstressed in shear and required shear stiffening, though none was provided;
(2)   the web of the column adjacent to the bottom flange of the haunch was overstressed in bearing and required horizontal compression stiffeners, though none was provided.

These omissions can be seen in Figs 10.6 and 10.8.

For this first test, bearing in mind the 'agricultural' nature of the design, fibre-reinforced cement sheets were used as roof cladding. Purlins for all of the tests were light gauge steel zed-sections fixed according to the manufacturer's recommendations. The purlins used

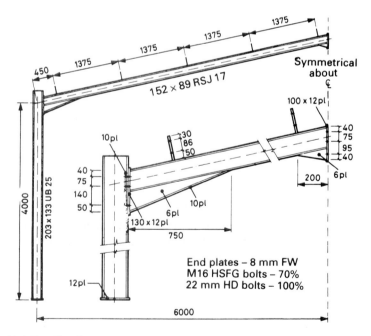

Fig. 10.6 Details of Frame 1.

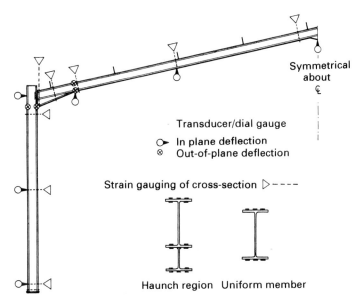

Fig. 10.7 Instrumentation of Frame 1.

were deliberately made approximately two sizes larger than the minimum indicated by the design calculations in order to ensure that the purlins did not fail before the frames. It may be noted that, in view of the relatively stocky members (RSJs) used in this particular test frame, the design did not include fly braces.

All cross-sections that required a detailed examination were fully strain gauged, particularly in the haunched region as shown in Fig. 10.7. This figure also shows locations of the linear deflection transducers. In addition, critical regions where yield was anticipated were coated with a brittle lacquer so that the spread of plasticity during the test could be monitored and recorded photographically.

As this was the first frame to be tested, considerable caution was exercised until the various control and measuring systems were proven. This meant initially loading the frame with relatively low intensities of load relative to the estimated failure load until the investigators were satisfied that all of the instrumentation was performing satisfactorily. Several preliminary tests were carried out before the loading was increased until the frame failed.

Failure of the frame was due primarily to a bearing/buckling failure of the column flange local to the bottom flange of the haunch, followed by the formation of a shear hinge in the column web, as shown in Fig. 10.8. The flaking of the brittle lacquer in the web is clearly visible

Fig. 10.8 Premature failure of Frame 1 at top of column.

and indicated the extent of the yielded zones. These primary modes of failure induced a secondary mode, i.e. lateral-torsional buckling of the column member. At this stage, the test was terminated in order not to overstrain the test assembly.

10.4 Theoretical analysis of Frame 1

Two theoretical analyses were carried out in comparison with the full-scale test. The first of these was an extremely detailed finite element analysis using eight noded isoparametric shell elements with second-order elastic-plastic capability. This is described in more detail in References 10.2 and 10.4 and will not be repeated here.

The second theoretical analysis was a second-order, elastic-plastic plane frame analysis as outlined in Chapter 3.

10.5 Behaviour of the first test frame

The finite element analysis for the first frame was undertaken in 30 displacement steps, assuming no previous load history. It finally 'failed' at a total load of 77.82 kN which is equivalent to a uniformly distributed roof load of 1.297 kN/m². Figure 10.9 shows both experimental and theoretical load-deflection curves for the vertical deflection at the apex. The theoretical predictions given by a simple second-order plane frame analysis were based on the measured full plastic moments of the members, and a knowledge of the base moments and rotations observed during the test. Curve (b), which incorporates a shear hinge, assumes that, when the web of the column behind the haunch had fully yielded in shear, a hinge was formed which was directly analogous to a plastic hinge, though having a lower moment of resistance than the plastic resistance of the column member. Figure 10.10 shows the bending moment diagrams corresponding to the test results, the finite element analysis and the simple analysis (with shear hinge) for the frame just prior to failure.

The following observations arise from the finite element analysis:
All the physical evidence suggests that this gave a faithful account of the failure history of the tested frame.

(1)   The extensive compression yielding in the column webs immediately behind the inclined compression flange of the haunch started when the load was approximately 0.55 kN/m². Above this load,

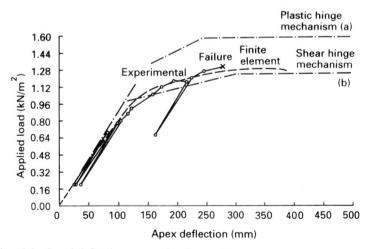

Fig. 10.9   Load-deflection curves for Frame 1.

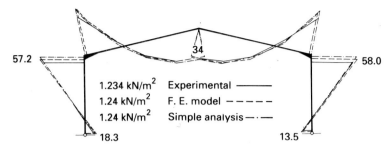

57.2   58.0

34

1.234 kN/m^2 Experimental ———
1.24 kN/m^2 F. E. model – – – – –
1.24 kN/m^2 Simple analysis —·—

18.3   13.5

Fig. 10.10   Bending moment distribution near failure of Frame 1.

these local areas of column webs showed signs of crushing and buckling, due to the absence of compression stiffeners.

(2)   As the geometric imperfections and material properties of the members were not uniform throughout the frame, the spread of plasticity on the weaker side caused the frame to sway in a non-symmetrical manner, though this trend reversed as failure was approached.

(3)   Shear yielding commenced in the web panel of the column head at a load of 0.75 kN/m^2. Shear hinges formed in the two column heads at loads of 1.05 kN/m^2 and 1.08 kN/m^2 respectively. This is reflected by the load-deflection curve in Fig. 10.9 where the slope of the curve changes noticeably in this region of load. The delay in the formation of the second shear hinge was due to the preceding non-symmetrical sway movement of the frame.

(4)   The development of a mechanism was curtailed because the initial out-of-straightness of the rafter induced lateral bending of the haunched regions and eventually twisting of the column.

As noted, the test frame failed prematurely because of local web crushing/buckling and the formation of a shear hinge in the column head. The primary mode of failure was followed by lateral-torsional buckling in the column member itself. This resulted in a significant reduction in ultimate design capacity of the frame. The calculated failure of the column head is shown in Fig. 10.11 which displays a remarkable similarity to the physical failure shown in Fig. 10.8. The theoretical model clearly predicted both modes of failure. From the theoretical analysis, the yielded zones in various parts of the frame at failure can also be deduced, as shown in Fig. 10.12. The diagrams clearly emphasise the extensive shear yielding in the column web and the local compression failure of the column member adjacent to the bottom flange of the haunch.

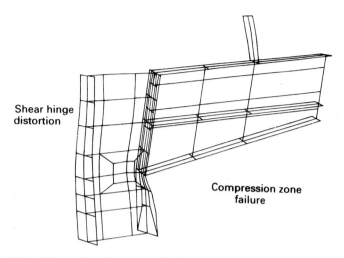

Fig. 10.11   Theoretical distortion of column head in Frame 1.

It may be noted that the stress distribution in the inner flange of the column at failure is highly non-uniform. A web bearing/buckling stiffener in the column is beneficial in reducing the effects of rolling tolerances and column web eccentricity which contribute to this effect.

The two load-deflection curves (a) and (b) in Fig. 10.9 obtained using plane frame analysis emphasise the influence of the unwelcome 'shear hinge'. Conventional plastic theory predicts a failure load of $1.58 \, \mathrm{kN/m^2}$ as shown by curve (a), whereas taking into account the reduced movement capacity of the shear hinge reduces the failure load to

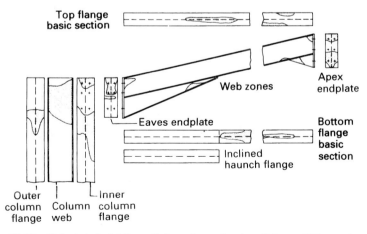

Fig. 10.12   Calculated yielding of the column head at failure of Frame 1.

1.25 kN/m² as shown by curve (b). Clearly, curve (b) is very much closer to the experimental results which obviously demonstrate the deleterious effect of the shear hinge.

Evidently it is important to carry out design checks on both the shear and bearing stresses in the haunch region and to incorporate the necessary stiffeners when the design checks indicate overstress. The tests reported here demonstrate that failure to do this may result in a significant reduction in the strength of the structure.

10.6   Details of Frame 2

The fabricated details of the second test frame are shown in Fig. 10.13. This frame was also designed as an agricultural building to BS 5502 with a total vertical load (dead plus imposed) of 0.67 kN/m² and a load factor of 1.7 giving a factored vertical load of 1.14 kN/m². However, this frame was also designed to resist a factored side load from stored grain of 40 kN acting horizontally at a height of 2.5 m above the base on one side only. This load tends to dominate the design and results in a significantly heavier frame.

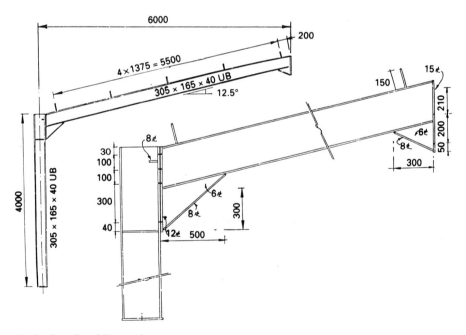

Fig. 10.13   Details of Frame 2.

Furthermore, the sponsors of the project specifically requested that the point of application of this side load should not be braced in the first instance as agricultural buildings are frequently built to accommodate such a load though with no bracing at the load point. The investigators thought that it would be interesting to comply with this request.

Examination of the haunch arrangement suggests that the toe angle of the haunch is rather large and a bearing stiffener should probably have been provided in the rafter. This factor does not appear to have influenced the results of the tests.

For this test, the roof cladding was a 32 mm deep profiled steel sheet supported on Z purlins at 1.375 m centres. Again, there was no shear stiffener behind the eaves connection and no fly braces. As shown in Fig. 10.14, however, there was a compression stiffener in line with the bottom flange of the haunch. Detailed calculations show that, strictly speaking, a shear stiffener should have been provided but the requirement was marginal and certainly not as critical as in Frame 1.

Fig. 10.14   Eaves detail of Frame 2.

Table 10.1 Test sequence for Frame 2.

Load case	Load applied	Failure mode (if applicable)	Maximum load applied
1	vertical only	elastic range only	1.63 kN/m^2
2	side load only	elastic range only	80 kN
3	combined, jack on right	lateral buckling of stanchion	load factor 3.16
4	combined, jack on left	local buckling at load point	load factor 2.82
5	combined, jack on left	plastic collapse mechanism	load factor 3.33

10.7   Test sequence for Frame 2

The presence of the side load allowed a rather more comprehensive test series than was possible with Frame 1 which was loaded by vertical load only. The test series was further increased when premature local failures were repaired and reinforced before testing to failure in an alternative mode. The resulting test sequence is summarised in Table 10.1.

10.8   Theoretical analysis of Frame 2

Although both detailed finite element analysis and relatively simple plane frame analyses were also available for this test, it is generally sufficient to discuss the test results primarily with reference to elastic-plastic plane frame analyses.

10.8.1   Load case 1 (vertical load only)

The experimental and theoretical bending moment diagrams at the maximum applied load are shown in Fig. 10.15. This agreement is close, confirming that the load application and strain measuring equipment were in good order.

10.8.2   Load case 2 (side load only)

The experimental and theoretical bending moment diagrams at the maximum load applied are shown in Fig. 10.16. There are small differences between the two, possibly reflecting local yielding at the load point, but the agreement is acceptable. The relatively large bending moments caused by the side load (stored materials) should be noted.

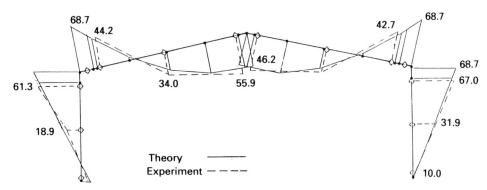

Fig. 10.15 Bending moment diagram (kNm) for load case 1.

10.8.3   Load case 3 (combined loading)

For this load case, both side load and vertical load were applied together. The bending moment diagrams at an intermediate stage of loading (1.20 kN/m² vertical + 42.0 kN horizontal) are shown in Fig. 10.17. Evidently, the redistribution of the high bending moments around the load point and the right-hand eaves are more marked here. This could be a combination of local yielding at the point of application of the load and movement in the bolted connection at the eaves. The shape of the bending moment diagram should also be noted with high and almost uniform moment in the upper part of the right-hand column.

This test was terminated when the right-hand column failed in lateral torsional buckling at a load factor of 3.16 (2.12 kN/m² vertical + 74.4 kN horizontal).

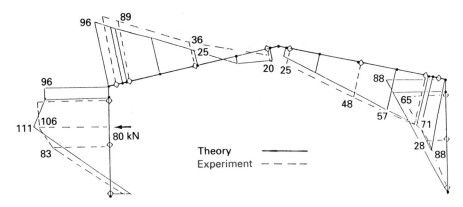

Fig. 10.16 Bending moment diagram (kNm) for load case 2.

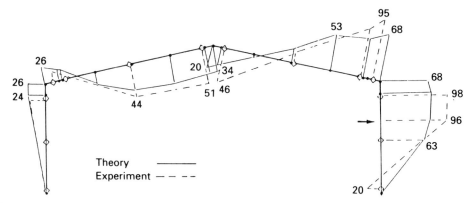

Fig. 10.17 Bending moment diagram (kNm) for load case 3.

10.8.4 Load case 4 (combined loading)

After the termination of the test to failure described for load case 3, it was observed that, apart from the right-hand column, the remainder of the frame was relatively undamaged. The shape of the bending moment diagram indicated that the left-hand half of the frame had been relatively lowly stressed. It was therefore decided that additional information could be obtained by continuing the test series.

Accordingly, the buckled column was taken out of the frame, straightened and replaced. The load system was then transferred to the undamaged left-hand column. A fly brace was introduced at the load point in order to prevent torsional movement.

When the combined loads were then increased together, a local buckling failure took place at the load point at a load factor of 2.82. This is, of course, a lower load level than was achieved in load case 3 with an unrestrained column.

Neither the bending moment diagram nor other measurements taken during the test gave any indication of why this should have occurred. Two possible reasons can be conjectured. Either the loading jack was misaligned or residual stresses locked into the frame as a result of load case 3 accelerated the failure.

10.8.5 Load case 5 (combined loading)

After completing load case 4, some further remedial work was under-taken. The local buckle was straightened and the buckled region

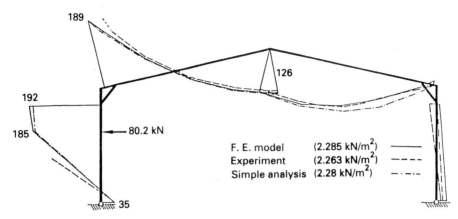

Fig. 10.18 Bending moment diagram (kNm) at collapse.

heavily stiffened. Then the frame was again tested to failure. This time a plastic collapse mechanism was achieved with a load factor of 3.33. The bending moment diagram at collapse is shown in Fig. 10.18.

The load-deflection curves for this final test to failure are shown in Fig. 10.19. After the introduction of the necessary additional bracing and strengthening, a ductile plastic collapse was obtained which could be adequately predicted by the usual analytical techniques.

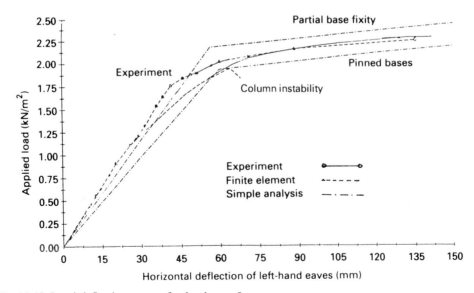

Fig. 10.19 Load deflection curves for load case 5.

10.9   Conclusions from the tests on Frame 2

The tests on Frame 2 have been described in some detail because they are considered to be extremely instructive. Experience of structural failures indicates that it is generally the details which fail. Under-design of the primary members is rarely a factor and experience with the large-scale testing of steel structures confirms this view. In order to avoid premature failure, it is necessary to pay particular attention to what are often considered to be the less important aspects of the design such as joint details and bracing.

Once the minimum amount of remedial work had been carried out, namely a fly brace and web stiffener at the load point, the relatively stocky frame was stable up to plastic collapse without any fly braces in the remainder of the structure. With agricultural buildings of small span, there is evidently some scope for judgement regarding the provision of bracing members. It is recommended that point loads of the order of magnitude of the crop load in this example should always be braced and stiffened. Fly braces should be introduced at points where rotational restraint is required except where RSJs or stocky Universal Beams at the smallest end of the range are used.

10.10 Acknowledgement

The testing described in this chapter was carried out at the University of Salford under the supervision of the first author in collaboration with Dr L. J. Morris of the University of Manchester. The generous financial support given by SERC, MAFF, CSC (UK) Ltd, IDC Ltd, Allott and Lomax and Robert Watson & Co is gratefully acknowledged. Gratitude is also expressed to Metal Sections Ltd for the supply of purlins and TAC and PMF for the supply of sheeting.

References

10.1   Engel P. The testing and analysis of pitched roof portal frames. PhD thesis, University of Salford, July 1990.

10.2   Davies J M *et al*. Realistic modelling of steel portal frame behaviour. The Structural Engineer, Vol. 68, No. 1, 1990, pp. 1–6.

10.3   British Standards Institution. BS 5502 Buildings and structures for agriculture: Part 20: Code of practice for general design considerations.

10.4   Liu T C H. Theoretical modelling of steel portal frame behaviour, PhD thesis, University of Manchester, October 1988.

Index